Nelson Science & Technology

Skills Handbook

Program Consultant
Marietta (Mars) Bloch

Authors
Nancy Dalgarno Alldred
Stephen Haberer

Contributing Writers
Paul Podesta
Mary Jane Vowles

Nelson
Thomson Learning™

Australia • Canada • Denmark • Japan • Mexico • New Zealand • Philippines
Puerto Rico • Singapore • South Africa • Spain • United Kingdom • United States

Contents

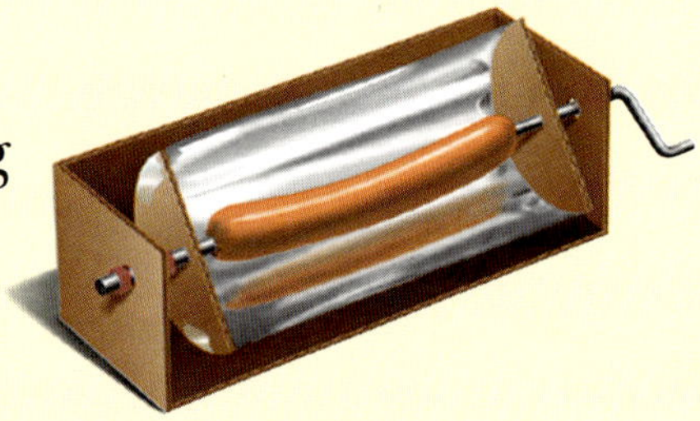

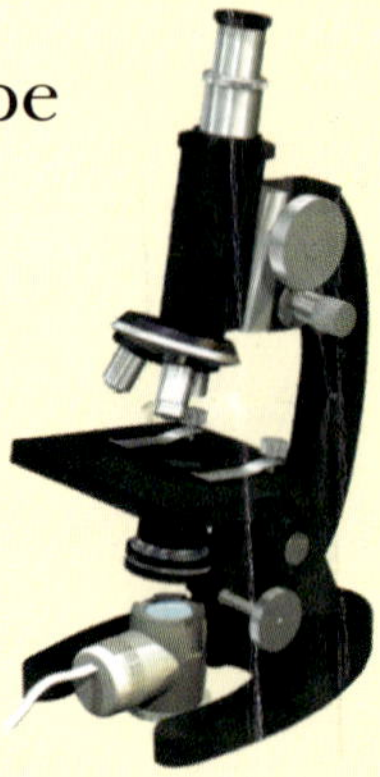

Important safety information

1 Safety in Science & Technology

Safety Conventions in *Nelson Science & Technology 7/8*

The investigations in *Nelson Science & Technology 7/8* are challenging, interesting, and safe. However, accidents can happen. In all of the investigations, potential hazards are identified by a caution symbol and described in red type (**Figure 1**).

Figure 1

Be careful when cutting. Always cut away from your body.

Always read the cautions carefully and make sure you understand what they mean before you proceed. If you are in doubt about anything, ask someone who knows (i.e., your teacher or a parent).

You should also look for caution symbols on the materials and products you use in your investigations. The symbols in **Figures 2 and 3** are often found on the labels of hazardous products.

Hazardous Household Product Symbols (HHPS)

You are probably familiar with the warning symbols in **Figure 2**. They appear on a number of products that are common in most households. These warning symbols indicate exactly why and to what degree a product is dangerous.

Workplace Hazardous Materials Information System (WHMIS) Symbols

The Workplace Hazardous Materials Information System (WHMIS) symbols in **Figure 3** were developed to standardize the labelling of dangerous materials used in all workplaces, including schools.

Preventing Accidents

Most accidents in the Science & Technology classroom are caused by carelessness. Following these general procedures can help you prevent them.

- Pay careful attention to instructions and take your work seriously.
- Be prepared for every investigation.

Figure 2
Hazardous Household Product Symbols (HHPS)

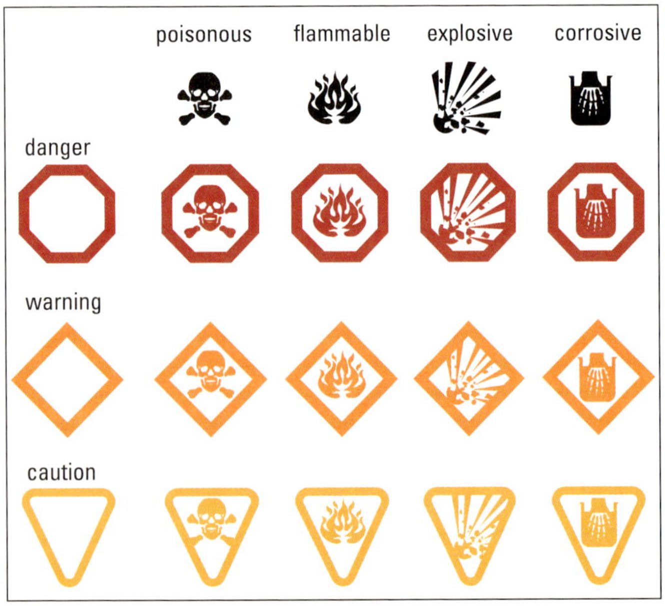

Figure 3
Workplace Hazardous Materials Information System (WHMIS) Symbols

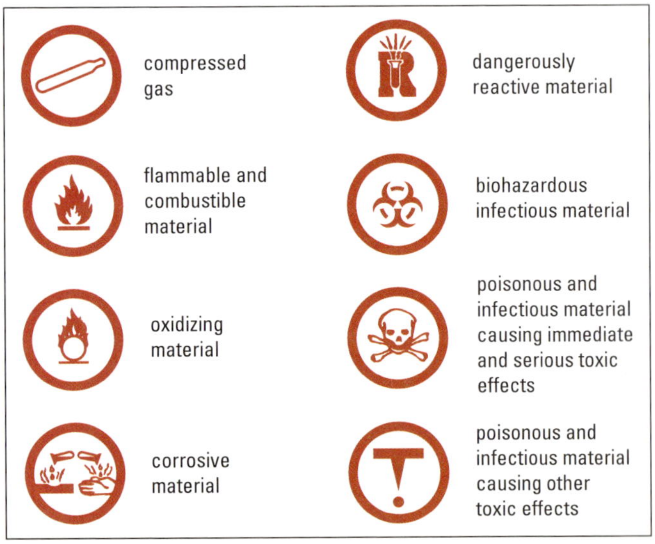

- Keep your working space clean and organized.
- Tie back long hair and don't wear loose clothing or jewellery.
- Measure and mix chemicals correctly.
- Handle hot equipment with care.
- Do not apply too much pressure to glass equipment (including microscope slides, cover slips, and thermometers).
- Use all tools, especially sharp tools, with proper care.
- Wear safety goggles.

Getting Off to a Safe Start

1. Learn the location and proper use of safety equipment, such as:
   - safety goggles
   - protective aprons
   - heat-resistant gloves
   - eye-wash station
   - broken-glass container
   - first-aid kit
   - fire extinguishers
   - fire blankets
   - water station

   Locate the nearest fire alarm.

   Know the procedures to follow in case an accident does occur.
2. Inform your teacher of any allergies or medical conditions that you may have. Do not wear contact lenses when conducting investigations—if a foreign substance became trapped beneath the lens, it would be difficult to remove it.
3. Read the procedure of any investigation carefully before you start. If there is anything you do not understand, ask your teacher to explain. Check all materials for warning labels. Make sure all tools and equipment are in good working condition and are properly secured. Clear your work area of all materials except those you will use in the investigation. If you are designing your own investigation, have your teacher check your design before you proceed.
4. Wear safety goggles and appropriate protective clothing, and tie back long hair (**Figure 4**). Remove jewellery and neckties. Wear closed shoes, not open sandals. Beware of loose clothing.
5. Return all equipment and tools to the appropriate storage facility.
6. Only use power tools if a qualified adult has trained you in their safe use.
7. Never work alone.

Figure 4

a Ready to conduct a safe inquiry investigation

b Ready to conduct a safe design investigation

Working with Chemicals

1. Do not taste, touch, or smell anything unless you are asked to do so by your teacher. Do not chew gum, eat, or drink in the Science & Technology classroom.
2. Be aware of where the Material Safety Data Sheet (MSDS) manual is kept. Know any relevant MSDS information for the chemicals you are working with.
3. Label all containers. When taking something from a bottle or other

container, double-check the label to be sure you are taking exactly what you need.

4. If any part of your body comes in contact with a chemical or specimen, wash the area immediately and thoroughly with water. If your eyes are affected, do not touch them but wash them immediately and continuously with cool water, moving from the bridge of the nose toward the outside corner of the eye, for at least 15 min. Inform your teacher.
5. Handle all chemicals carefully. When you are instructed to smell a chemical, wave the vapour toward your nose (**Figure 5**). This way, you can smell the substance without inhaling too much into your lungs. Never put your nose close to a chemical.

Figure 5
Safe smelling

6. Place test tubes in a rack before pouring liquids into them. If you must hold a test tube, tilt it away from you (and others) before pouring in a liquid (**Figure 6**).

Figure 6
Safe pouring

7. Clean up any spilled or dropped materials immediately, following instructions given by your teacher.
8. Do not return unused chemicals to the original containers, and do not pour them down the drain. Dispose of chemicals as instructed by your teacher.
9. Ensure the area is well ventilated when using chemicals.
10. Clean all equipment thoroughly after use.

Working with Heat

1. Whenever possible, use electric hot plates for heating materials. Use a flame only if instructed to do so. If a Bunsen burner (**Figure 7**) is used in your Science & Technology classroom, make sure you follow the procedures listed below.
   - Obtain instructions from your teacher on the proper method of lighting and adjusting the Bunsen burner.
   - Do not heat a flammable material (for example, alcohol) over a Bunsen burner. Make sure there are no flammable materials nearby.
   - Do not leave a lighted Bunsen burner unattended.
   - Always turn off the gas at the valve, not at the base of the Bunsen burner.

Figure 7
Bunsen burner

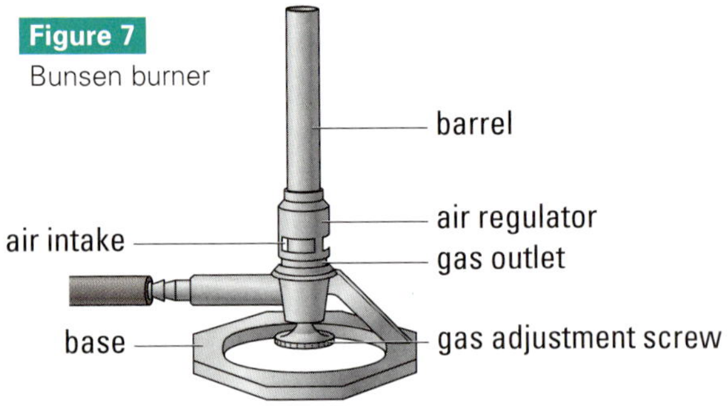

2. When heating liquids in glass containers, make sure you use clean Pyrex or Kimax. Do not use broken or cracked glassware. Never allow a container to boil dry.
3. When heating a test tube over a flame, use a test-tube holder. Hold the test tube at an angle, with the opening facing away from

you and others (**Figure 8**). Heat the upper half of the liquid first, then move it gently in the flame, to distribute the heat evenly.

Figure 8
Heating using a test-tube

4. Be careful when handling hot objects and objects that might be hot. Hot plates can take up to 60 min to cool completely. Test that they are cool enough to move by touching first with a damp paper towel. If you hear sizzling or see steam, wait a little longer! If you burn yourself, immediately apply cold water or ice, and inform your teacher.
5. Always have an inspected and certified fire extinguisher at hand.

Working with Light

1. Never look directly into a light source (including the Sun). Do not use magnifying lenses to observe light directly.
2. Use mirrors with care. If a mirror is broken, dispose of glass in the appropriate receptacle.
3. Since ray boxes are used in the dark, be careful when moving from place to place. Allow your eyes to become accustomed to the dark prior to beginning an activity.

Working with Sharp Tools

All tools must be used below waist level. Always wear safety goggles.

1. Saws (**Figure 9**):
   - Always use the correct type of mitre box to match the saw you are using. Be sure that the mitre box is secured to the work surface.
   - Make sure the saw blade is properly in place (the teeth of the saw should be angled away from your body).

Figure 9
The safe use of a saw

2. Drills:
   - Change drill bits carefully. To reduce breakage, always use high quality drill bits.
   - Drill on a solid, flat, secured surface. Be sure to secure the object you are drilling.
   - Always drill downwards.
3. Paper Drills:
   - Secure the cutting end.
   - Drill on a solid, flat, secured surface.
   - Use light pressure only.
4. Utility Knife:
   - Always use a metal safety ruler.
   - Cut on a solid, flat, secured surface.
   - Keep work area clear.
   - Always cut by pulling the knife toward you. Keep your other hand well out of the way. Work carefully and slowly. Use light pressure. Thicker materials may need several cuts (not more pressure).
   - Always retract the blade when finished.
5. Snips and Scissors:
   - Always carry these by the blade end.
   - Place the material being cut close to the pivot point so that it is easier to cut.
   - Snip materials so that they fall down onto the working surface.
   - When snips are not in use, secure the safety catch.

6. Hammer:
   - Hammer only the nail head. Never bang two hammers together.
   - Work on a solid, flat, secured surface.
7. Screwdriver:
   - Use a screwdriver only for its intended purpose, turning screws. Do not use as a punch, wedge, etc.

Working with Adhesives and Fasteners

1. Always read the label to ensure that what you are using is not toxic.
2. Follow directions for safe use. Be aware of first aid precautions and treatment.
3. Always use glue guns and adhesives in a well-ventilated area.
4. Glue Guns (**Figure 10**):
   - Wear safety goggles.
   - Always use a glue gun in a low traffic area of the room, near an electrical outlet. Ensure that glue gun cords are out of the work area.
   - Your work surface should be protected.
   - Always glue down to the object. Gluing should take place at the level of the work surface area.
   - To avoid burning fingers, always use a stick to position an object being glued. Never use a foreign object to force the glue through the barrel of the glue gun. Have cool water available in case of burns. Immerse burn for at least 10 min and inform your teacher.

Figure 10
Using a glue gun safely

5. Fasteners: Always use caution when putting metal fasteners in or through various materials. Whenever possible, precut the hole.

Other Hazards

1. Keep water and wet hands away from electrical cords, plugs, and sockets. Always unplug electrical cords by pulling on the plug, not the cord. Report any frayed cords or damaged outlets to your teacher. Make sure electrical cords are not placed where someone could trip over them.
2. Place broken and waste glass in the specially marked containers.
3. Follow your teacher's instructions when disposing of waste materials.
4. Report to your teacher all accidents and injuries (no matter how minor), broken equipment, damaged or defective facilities, and suspicious-looking chemicals.
5. Wash your hands thoroughly, using soap and warm water, after every inquiry or design investigation. This practice is especially important when you handle chemicals, biological specimens, and microorganisms.
6. Store chemicals and solvents in a secure, well-ventilated area.
7. Store tools and equipment in a safe place after use.

Try This — Safety Posters

In groups, create safety posters for your classroom. These may include a map of the route your class should follow when the fire alarm sounds, a map of where the safety materials (fire extinguisher, first aid kit, etc.) are located in your classroom, information about the safe use of a specific tool, or an illustrated poster outlining specific safety rules, etc.

2A Process of Scientific Inquiry

Science is knowledge of the world around us. It's also a way of learning about the world around us by observing things, asking questions, proposing answers, and testing those answers. Science is also about sharing information that encourages others to discover more. All scientists use a similar process to find answers to their questions. Let's look at an example of how this process is used before examining the specific steps.

Using the Process of Scientific Inquiry

From Curiosity to a Testable Hypothesis

After observing plants for some time, you may observe that they grow upward (**Figure 1**).

Figure 1

Why do plants grow upward?

If you're curious, you may ask yourself "Why?" You can probably think of some possible answers. For example, maybe plants grow upward because they grow toward the light. This is a hypothesis. From your hypothesis, you can make a prediction that you can test through experimentation. You could test the following prediction: "As the angle of light shining on plants steadily increases, the angle of the plants' growth should also steadily increase."

The Investigation

Once you have a prediction, you can design a test. You can think of your prediction as a statement of cause and effect. The effect you identified is the angle of plant growth. The cause you identified is the angle of light shining on the plant. Your investigation should test whether the cause will produce the effect you predict. The cause and effect can change or vary—so they are known as variables. To plan a careful, accurate investigation, you need to control the cause variable as much as possible. For example, you could investigate the growth of five identical plants, ensuring that all growing conditions—water, nutrients, temperature, kind of light, etc.—are the same (**Figure 2**). You would then change one variable, the angle of light shone on one plant, and measure the results.

Figure 2

Does changing the angle of light change the angle of growth?

The Results

When you design your investigation, you need to think about what you are going to measure and how you are going to record your results. For instance, you might create a data table like the one in **Table 1**.

Table 1 **Record of Angle of Growth**

Angle of Light	Day 1	Day 3	Day 5	Day 7	Day 9	Day 11
Sample A (0º)						
Sample B (30º)						
Sample C (45º)						
Sample D (60º)						
Sample E (90º)						

Once you have completed the investigation, you will have to analyze your results to see whether or not they support your hypothesis. You may create graphs based on the data you collect that will help you make sense of your results. Ultimately, your conclusions will be communicated to others in a report.

Unfortunately, you can never be sure that all possible causes of the effects have been controlled. This means that you can never be absolutely sure that your conclusions are true. However, the more closely the results match the prediction, the more confident you can be that your hypothesis was sound. If you communicate your procedure and conclusions clearly, the experiment can be repeated and the hypothesis re-evaluated. New questions may arise, and new evidence may be gathered.

The Steps in the Process of Scientific Inquiry

It is important that you follow this process and use the related skills whenever you are asked to design and conduct an experiment, if you expect to find reliable answers to the questions you pose. You can use the flow chart in **Figure 3** as a checklist. You can also refer to the more detailed sections in the *Skills Handbook* that deal with each of the specific skills necessary for each part of the process.

Try This — Detective as Scientist

Read a mystery story and describe which of the steps in the inquiry process were used by the detective to solve the crime. Use the flowchart in **Figure 3** as a checklist. The following list of questions may help you:

1. What question did the detective have to answer?
2. How did the detective reach a hypothesis? Were observation, research, or prediction used?
3. How was the hypothesis tested? Were variables controlled? Were clues observed? Were scientific instruments used?
4. How were the results of the investigation analyzed? How were observations recorded? What trends or patterns emerged? What conclusions were drawn? What new questions emerged?
5. How was the conclusion communicated?

1 Asking a Question

- Ask a question that interests you or express an idea that can be tested.

2 Making a Hypothesis

- Research information and previous discoveries that might help you answer your question.
- Develop an educated guess that answers your question. This is your hypothesis.
- Make a prediction based on your hypothesis and state it as a cause-effect relationship.

3 Designing the Experiment

- Identify all your variables.
- Decide what materials and equipment you will need to perform your experiment.
- Write a procedure that explains how you will conduct your experiment. Be sure to take safety into account.
- Draw a labelled diagram that visually explains your procedure and the material and equipment you will use in your experiment.
- Create a rough draft of tables for recording your data.

4 Conducting the Experiment

- Follow the steps in the procedure carefully and thoroughly.
- Use all equipment and materials safely, accurately, and with precision.
- Record the variable(s) you are measuring, manipulating and controlling.
- Remember to repeat your experiment at least three times. If you are collecting quantitative data, take an average. This increases the accuracy and reliability of your results.

Figure 3

The Steps in the Process of Scientific Inquiry

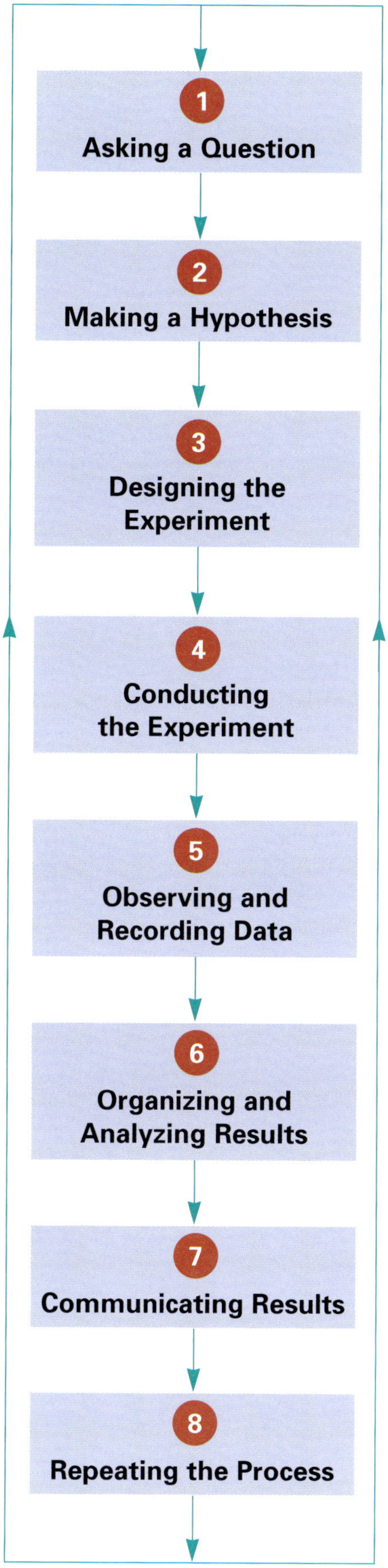

5 Observing and Recording Data

- Make careful notes of everything that you observe during the experiment.
- Record numerical data in tables.

6 Organizing and Analyzing Results

- When appropriate, create graphs to make better sense of data from your tables.
- Study all of your observations.
- Identify patterns and trends.
- Make a conclusion. Be sure that your results support, partially support, or reject your hypothesis.
- Develop an explanation for your conclusion.
- Apply your findings to your life today. Think about who will want to know about your discovery, how it will affect our lives, who it will benefit, and whether or not it could harm our world if used in a certain way.
- Reflect on your experiment. Explore any sources of experimental error in your process. Think about what changes you would make if you were to conduct this experiment in the future.

7 Communicating Results

- Write a report to summarize your investigation. In order to have their investigations repeated and validated, it is common for scientists to publish details of their research, results, and conclusions.

8 Repeating the Process

- If your experiment did not answer the question you initially asked, revise it and repeat it until the question is answered. Most scientists must complete an experiment many times before making important discoveries (e.g., a cure for cancer).

2B Asking a Question

Wondering About the World Around You

All scientific questions are asked by people who are curious about the world around them. Our curiosity is fed by observation, experience, and research. You may have noticed, for example, that balloons stick to walls if you rub them on your head (**Figure 1**). This observation might lead you to wonder about a number of things: Does a balloon stick better if you rub it more times? Does the length or texture of a person's hair affect how well a balloon sticks? Does it matter how inflated the balloon is or what colour it is?

Each of these questions is the basis for a sound scientific investigation. With a little rewording, it will become clear how these questions can be tested. Not every question that you will want an answer for *can* be tested: some are too general or too vague. Learning to ask questions that can be tested takes time and is a fundamental skill in scientific inquiry.

Figure 1
What affects how well the balloon sticks to the wall?

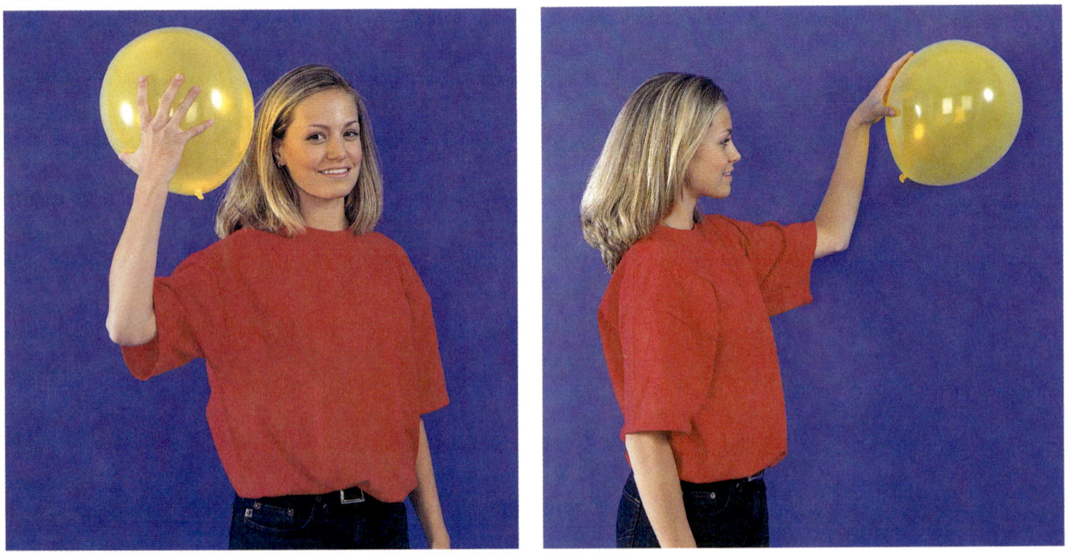

Questions About Cause and Effect

The first balloon-related question could be stated more usefully: "What is the effect of increasing the number of times that a balloon is rubbed on a person's head on the time that the balloon stays stuck to a wall?" You can probably already begin to plan an experiment that would answer this question. A testable question is often about cause-effect relationships. These questions often take the form: "What causes the change in variables?" and "What are the effects on a variable if we change another variable?" As you know, a variable is something that can change or vary in an investigation.

Variables and Questions You Can Test

Scientists call the cause variable the independent variable. This is the one thing in the experiment that you purposely change. For instance, increasing the number of times that a balloon is rubbed on a person's head is something you control: you are changing the independent variable.

The effect variable is called the dependent variable. This is what you measure in your experiment (e.g., time, distance) and "depends" on what variable you purposely change (independent variable). The amount of time the balloon stays on the wall is the dependent variable.

A scientific question that asks what happens to a dependent variable when we change the independent variable is a question you can test. Some examples appear in **Table 1**.

Table 1 **Testable Questions**

Question	Independent (cause) variable	Dependent (effect) variable
How do fertilizers and phosphates affect the productivity of algae?	fertilizers and phosphates	rate of algae growth
What happens to the volume of a liquid when it is heated?	applying heat	volume of the liquid
Do all transparent materials refract light the same amount?	kind of transparent material	amount of light refracted
Does stirring affect how quickly a solute dissolves in a solvent?	stirring the solvent	rate at which the solute dissolves

Try This What's the Question?

Have you ever kicked a ball during the winter? It almost feels like kicking a brick.

1. Place a tennis ball in a refrigerator overnight.
2. Reproduce **Table 2** for recording your observations.

Table 2

Temperature of ball	Height of ball (cm)	Trial 1	Trial 2	Trial 3	Average
cold	?	?	?	?	?
room temperature	?	?	?	?	?

3. Hold a metre stick vertically with the zero end on the floor.
4. Drop the cold tennis ball from the upper end of the metre stick (**Figure 2**).

Figure 2

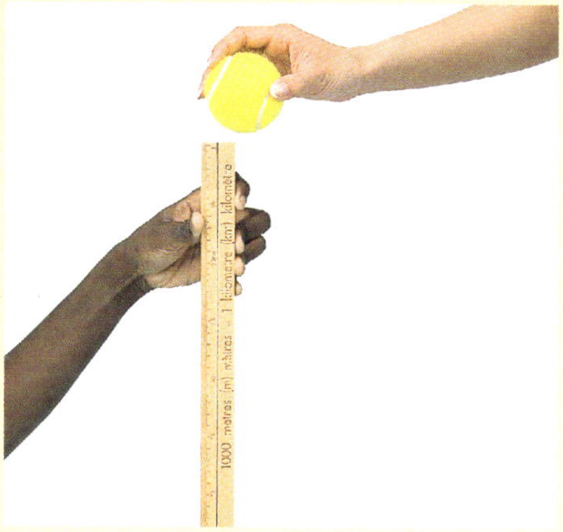

5. Measure the height that the tennis ball bounces back (**Figure 3**).

Figure 3

6. Do this three times (each time is called a test or a trial).
7. Repeat steps 3–6 with a ball that is at room temperature.
   - What question is being investigated in steps 1–6?
   - What question is being investigated in step 7?
   - What evidence is being gathered to answer the questions?

2C Predicting and Hypothesizing

The Hypothesis

A suggested answer or reason why one variable affects another in a certain way is called a hypothesis. Often you have some idea about this even as you ask your question. Having noticed, for instance, that a balloon sticks to the wall after you rub it in your hair, you might predict that rubbing it more will make it stick longer. You might be wrong, of course, but your prediction is probably based on past observations, on logic, and on bits of scientific theory you may remember. If you're really interested, you may even do some research based on what you already know. If you then pull everything you know together and express it, you would have a hypothesis. For instance, you might posit that the balloon sticks to the wall because it is attracted by static electricity, and that rubbing the balloon more produces a greater static electric charge on the balloon.

The Relationship Between Hypotheses and Predictions

Predictions and hypotheses go hand in hand (**Table 1**). The hypothesis is how you can explain a prediction. The prediction is what you test through your experiment. And, if the experiment confirms the prediction, you can have more confidence that your hypothesis is correct. Naturally, your experiment will not always confirm your prediction. In this case, you may need to re-evaluate your hypothesis and design a new experiment.

Try This — Answer These Questions

Support a plank with some books, so it forms an inclined plane (**Figure 1**). When you roll a jar down the plane, which will roll farther: an empty jar, or one filled with water? Write a hypothesis and a prediction to answer that question. Then carry out the experiment. How accurate was your prediction?

Now empty half of the water from the jar (**Figure 2**). How far will it roll? Write a hypothesis and prediction to answer the question, and try the experiment.

Do you need to change your hypothesis in either case?

Figure 1

Figure 2

Table 1 Sample Hypotheses and Predictions

Hypothesis (possible reason for cause-effect relationship)	Prediction: Possible cause (independent variable)	Prediction: Possible effect (dependent variable)
Candy contains sugar that is used for energy by germs in the mouth, and these germs produce an acid that decays the teeth.	As the amount of candy that a person eats increases...	...the number of tooth cavities increases.
A larger sail traps more air, which then provides a greater force to a boat.	As the size of the sail increases...	...the top speed of the sailboat increases.
Salt helps oxygen in the air combine with iron in the metal of a bicycle to form rust.	As the amount of salt on a road increases...	...the amount of rusting of the metal parts of a bicycle increases.

2D Identifying Variables and Controls

You may notice that milk has a different taste if it is left out on the table for more than 3 h. What additional factors, other than the time that it has been left on the table, do you think might affect the taste of milk? You have just identified variables!

Identifying Variables

Identifying variables is extremely important to the process of scientific inquiry: it allows you to ask good scientific questions, make predictions, and design a meaningful experiment. As you know, anything that can change in an experiment is a variable. Any variable that can be changed by the experimenter is called an independent variable. For example, you might predict that if a jar has a greater mass, then it will roll farther after it leaves a ramp. The mass of the jar is the independent (or cause) variable. There are other independent variables in this experiment, shown in **Figure 1 a–c.**

The effect of changing the independent variable is called the dependent variable. The distance that the jar travels is the dependent variable.

Figure 1

a The angle of the ramp is an independent variable.

b The size of the jar is an independent variable.

c The type of liquid in the jar is an independent variable.

Identifying Controls

Scientists attempt to test only one independent variable at a time. That way the scientist knows which cause produced the effect. Controls are used to eliminate the possibility of an unknown variable.

In the jar experiment, you need to control the independent variables. To isolate the mass of the jar as the cause, you must try to change only the mass. The angle of the ramp, the size of the jar, and the type of liquid you use must always be the same—only the amount of liquid can change. If you try the experiment, you may find that the results with a half-filled jar surprise you. As a half-filled jar rolls down the ramp, water sloshes back and forth, creating friction. The friction inside the jar slows it down, and it doesn't travel as far. Friction is another independent variable that needs to be controlled. The jar in **Figure 2** with no water in it isn't affected by this variable: it is the control.

Figure 2
The empty jar on the ramp is a control.

> ### Try This — Don't Get Burned!
>
> Do sunscreens really prevent sunburn?
>
> 1. State a related hypothesis and prediction.
> 2. Identify the dependent and independent variables in your experiment.
> 3. What would your controls be?

Why Do Scientists Use Controls?

The following experiment shows why scientists use a control. A student notices spots of rust on the frame of her bike after riding all winter. The student hypothesizes that the rust was caused by the salt on the roads. But is salt the only variable (**Figure 3a**)? It is possible that the rusting was not caused by the salt alone. For example, the roads were also wet a lot of the winter. By comparing the amount of rusting with and without salt, you can determine just how much rusting is caused by the salt (**Figure 3b**).

Figure 3

a Your prediction: As metal is exposed to more salt, more rusting will occur.

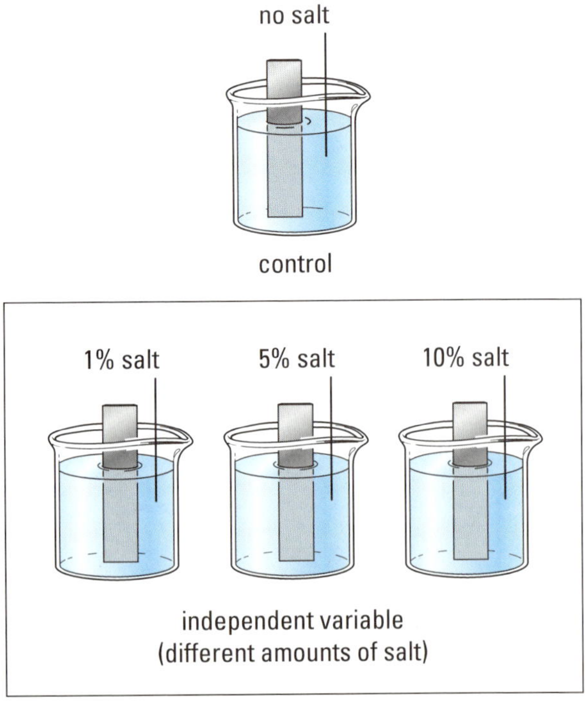

b The experiment: Metal placed in different solutions

2E Designing an Inquiry Investigation

Once you have asked a testable question and developed a hypothesis, you can design an investigation that will test that hypothesis. To design an inquiry investigation, you need to think of all the steps in the process of scientific inquiry outlined in section 2A. You need to visualize the investigation from beginning to end and plan for everything you'll need and every step you'll take.

Variables

Let's imagine that your question is: "Does mass affect the swing of a pendulum?" Your prediction may be: "As the mass of the pendulum increases, the number of swings per second will increase." The first step in designing your inquiry investigation is to identify your variables. Remember that an independent variable is the possible cause and a dependent variable is what the effect will be. You know you will be using different masses in your experiment (which you will have to record), and you will be measuring the number of swings per second. But you must also identify other possible independent variables so that you can control them. As you picture yourself testing the pendulum (**Figure 1**), you can probably list some of these possible causes. For instance, the length of the string, the length of the arc, the materials from which the pendulum is made—all of these may affect the swing. You must design your experiment so that these variables are kept constant.

Observations

You also need to think carefully about what you are going to measure—in this case, mass and number of swings of the pendulum. What mass are you going to start with? And how are you going to increase the mass? What unit of time are you going to use? And how long will you observe the pendulum in each case? Then you need to design a table to record your data (**Table 1**).

Table 1

The effect of mass on number of swings per second (based on 15 s of observation)	
Mass	**Number of swings per second**
1 g	?
2 g	?
3 g	?
4 g	?

Equipment and Materials

Obviously, you must also decide what materials and equipment you will use (**Figure 2**). It is useful to create a labelled diagram that illustrates the materials and equipment you will need, and the procedure you are going to use.

Figure 1
What are the independent variables?

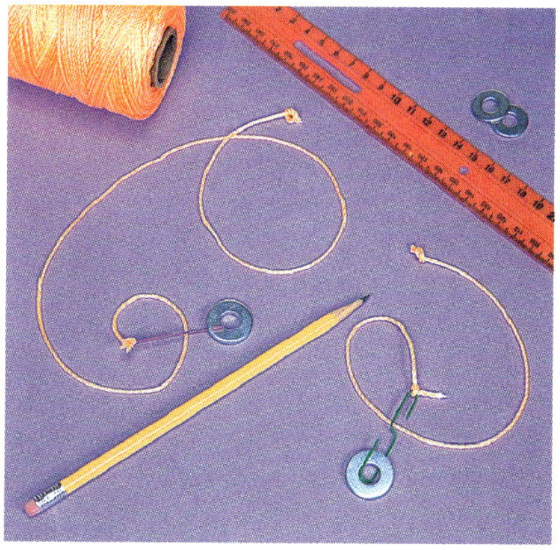

Figure 2
What materials will you need?

The Procedure

This is an essential component of an experimental design. Anyone who is interested in learning about your experiment needs to be able to understand how it was performed, so that it can be duplicated exactly (**Figure 3**). Therefore, it is important that you be able to write an experimental procedure clearly, concisely, and accurately. When writing a procedure, you should use:

- numbered steps
- passive voice (avoid using pronouns)
- past tense

Figure 3

Record every step you will take.

For example, the first two steps of a procedure could look like **Figure 4**.

Don't forget to consider all possible safety issues!

Figure 4

Procedure

1. The experiment was set up as shown in the diagram.
2. In each case, the pendulum was observed for 15 s.

Try This Write a Procedure

Think about how you would conduct the pendulum experiment. Write a procedure for this experiment, using the format explained above.

3A The Problem-Solving Cycle

The major goal of technology is to improve the world by creating products that satisfy human needs. For example, while scientists try to explain why and how the mosquito finds its prey, technologists try to create products to protect us from these annoying pests (**Figure 1**).

Figure 1

a Why does the mosquito bite?

b How can we stop the mosquito from biting?

For example, consider automobiles, computers, plastics, or the process for making decaffeinated soft drinks. Most of us are most creative when we're trying to satisfy a need. Imagine, for instance, that you and a few friends are alone at home. You're hungry, but there's not a lot of food in the refrigerator that you're really interested in. You all agree that hotdogs would be acceptable, but the power is out. Your problem is to design a method of cooking hotdogs without electricity. Luckily, your friends are with you. Working together to brainstorm possible solutions and to identify the advantages and disadvantages will help you come up with a good solution to your problem (**Figure 2**).

Figure 2

Technologists use a process of design to solve problems. Let's look at an example of how this process is used before examining the specific steps.

Using the Process of Design

What Are You Designing?

Things are usually designed in response to needs and to try to improve the quality of life.

Creating the Solution

You decide to use the energy from the Sun to cook your hotdogs. You know that aluminum foil reflects heat, so you will use it to get the most from the Sun's heat. And you want to make sure the hotdog cooks evenly. You jot down ideas and sketch out possible designs, considering all the tools and materials that you will need. You talk through how to build your hotdog cooker and realize there are a couple of things you hadn't thought of. When you think you're ready to build, you gather everything you need and go to work. Together you have built a prototype hotdog-cooker that is designed to solve your problem (**Figure 3**).

Figure 3
The Sun cooks lunch.

What Have You Learned?

The first hotdog you cook may not have turned out perfectly. Maybe the wire wasn't turning freely, and you had trouble cooking your hotdog evenly. That first hotdog was a test of your prototype, and you discovered that your design could be improved. You need to back up a few steps and modify your prototype. Once you get it working right, you may want to show some of your other friends how to make a solar hotdog cooker. Communicating the steps of the process is a very important part of the process.

The Steps in the Process of Design

The steps in the process of design are very similar to the steps used in the process of scientific inquiry. The flow chart in **Figure 5** identifies those steps. Use it as a checklist when you are required to design and test a prototype you construct in your Science & Technology class this year.

Try This Design a Mould

It's your little sister's birthday, and she loves strawberry gelatine (**Figure 4**). You've bought the powder and are trying to find something to make it in. You'd like the end result to look fun and festive. All you have are three bowls, and none of them is the right size. You've used one of the bowls in the past, and it was hard to get the gelatine out once it was set.

- Write a problem statement, choose a solution, and plan a prototype for a jelly mould that you think would solve your problem.

Figure 4

Figure 5
The Steps in the Process of Design

1 Identifying a Problem

- Identify and record the problem.
- Identify any limits posed by the problem.
- Research possible solutions to the problem and record new ideas.
- Restate the problem so that it clearly states what needs to be solved.

2 Selecting the Best Alternative

- Identify the constraints (e.g., human resources, time, money, environmental issues).
- Develop the design criteria (e.g., cost, reliability, safety, size, materials).
- Brainstorm all possible solutions to the problem.
- Rate the possible solutions.
- Select the alternative with the highest rating.

3 Planning the Prototype

- Draw sketches and/or schematics of your prototype.
- List materials and tools that will be needed to construct the prototype and gather them.

4 Building the Prototype

- Refer to (1) "Safety in Science & Technology" for a review of safety precautions when using tools.
- Construct a working prototype.

5 Testing and Evaluating the Prototype

- Use the design criteria to develop a test.
- Test the prototype to evaluate its ability to solve the identified problem.
- Quantify and record your test results.

6 Communicating

- Create a Design Folder that describes your process and product, and communicates your evaluation of how well the product satisfied the design criteria.

7 Repeating the Cycle

- Revise and reconstruct the prototype in order to make it more effective.
- Repeat the process until an effective solution to the problem has been developed.

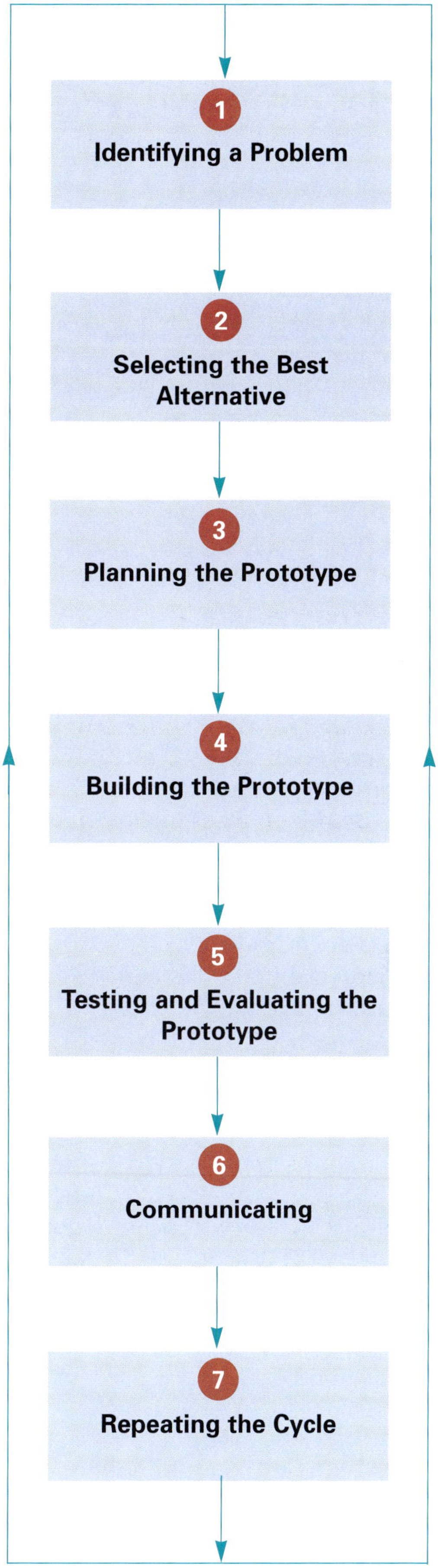

3B Identifying a Problem

Identify and Record the Problem

It is important that, as you start the design process, you understand and specify the nature of the problem. Start by writing down what you believe the problem is, based on the information you have gathered. Often the problem is the need of a client. The problem may be written as a question or a statement.

Let's assume that your neighbours have a need and that they are your clients. They have ten goldfish that they dearly love. Occasionally, they go away for week-long vacations. Your neighbours are worried about how to feed the fish. They would like you to design and build a device that drops a small quantity of goldfish food into the goldfish tank twice a day (**Figure 1**).

Figure 1

Goldfish require daily feeding.

Your stated problem could be:

A device is necessary to automatically feed ten goldfish two times a day for a period of one week.

Examine and Redefine the Problem

It is important to fully understand this problem statement and to redefine the problem so that it is as clear and objective as possible. The following steps may lead you to rephrase your problem statement:

- **Be sure you fully understand the terms stated in the problem.** You may have to clarify terms or make them more precise. For example, in the given problem it is not clear how much goldfish food is to be released—the amount of food should be specified. The problem could be restated:

  *A device is necessary to automatically feed ten goldfish **0.5g** of food, two times a day, for a period of one week.*

- **Identify limits posed by the problem as stated.** For instance, the problem may have been expressed by someone who already has an idea about the "best" solution. In the problem statement above, it is obvious that your clients think that an automatic device is necessary to feed the goldfish. It may, however, be possible to hire a person to come in and feed the fish. Therefore, the problem may be refined to say:

  *A **method** or device is necessary to feed ten goldfish 0.5g of food, two times a day, for a period of one week.*

- **Research possible solutions to the problem and record new ideas.** You may find an appropriate solution that already exists. Or you may get new ideas that will help you focus your problem-solving.

Try This: Listing Needs

Generate a list of 10–15 needs that could be met through technology. Categorize the needs under the headings in **Table 1**:

Table 1

Personal needs	Class needs	School needs	Community needs
?	?	?	?
?	?	?	?
?	?	?	?

3C Selecting the Best Alternative

Identifying Constraints

For every design problem there will be factors that limit or constrain a solution. If you are designing an automatic goldfish-feeder, for instance, your clients will probably give you a limited amount of money and time to complete the project. The money and time factors are limits or constraints.

General Constraints

General constraints that should be considered for most technological designs are listed below:

- **Natural Constraints:** Will the laws of nature let me do this?
- **Time Constraints:** How long will this take? Is this a major project involving years of development and testing or a relatively short endeavour?
- **Financial Constraints:** How much money do I have to pay for the development of this?
- **Material Constraints:** What materials and tools are available to complete this project?
- **Societal Constraints:** Is there actually a need or demand for this product?
- **Ethical Constraints:** Should we be doing this? Is this harmful for society in any way?
- **Environmental Constraints:** What impact will this have on the environment?

Evaluating these constraints will influence what solutions are available and affect the specific design criteria for a project (**Figure 1**).

Developing Design Criteria

Design criteria clearly outline what your product needs to achieve if it is to successfully solve the problem for which it was designed. After answering the questions in the list of general constraints you may come up with some of the design criteria for your project. For instance, your clients may be going on vacation in three weeks. In this case, one of your design criteria would be that the automatic goldfish-feeder must be complete in three weeks.

Other design criteria will be determined by the specific nature of the problem you are trying to solve. Reliability, size, appearance, strength, durability, safety, efficiency, and maintenance are just some of the factors that may be relevant to your design.

It is useful to brainstorm a list of criteria, judging how important each is to the success of the design. For an automatic goldfish-feeder, for instance, you might consider the following design criteria:

- **Time:** The goldfish-feeder must be complete in three weeks.
- **Money:** It must cost $120 or less.
- **Reliability:** It must be completely reliable, or the fish will die.
- **Ease of use:** It is of medium importance that the device be easy to use. It has to be set up correctly, but the owner can take time to ensure that it has been correctly set up before leaving the fish unattended.
- **Appearance:** It is of low importance for the device to be attractive since it is in use only when people are away.

Figure 1
The amount of money available, for instance, will affect what materials can be used and how many people you can hire.

Brainstorming Solutions

It is now time for you to brainstorm a list of possible solutions to your problem (**Figure 2**). This may involve coming up with an idea for a new product, a modification to a device that already exists, or a new or modified process.

Figure 2
When brainstorming, each and every suggestion should be recorded—don't stifle creativity by discouraging other people's ideas.

After you have completed your brainstorming exercise, vote for and list the best alternatives.

Rating the Alternatives

We now need a method for choosing the very best alternative. There are different ways of rating alternatives, but it is important you measure each alternative against your design criteria.

The problem-solving matrix in **Table 1** is one way you can rate alternatives. **Table 2** provides an example of how the matrix can be used to rank alternatives to the automatic fish-feeder challenge. In this example, the lower the number, the worse the rating. Note that the criteria are measured according to different scales, depending on the relative importance of each criterion.

Table 1 **Generic Problem Solving Matrix**

	Design Criteria					
	Criterion A	Criterion B	Criterion C			
Alternatives	rating scale	rating scale	rating scale			Totals
Alternative A	?	?	?	?	?	?
Alternative B	?	?	?	?	?	?
Alternative C	?	?	?	?	?	?

Table 2 **Problem Solving Matrix – Fish-Feeder Example**

	Design criteria					
	Time	Cost	Reliability	Ease of Use	Appearance	
Alternatives	(0-10)	(0-10)	(0-10)	(0-5)	(0-3)	Totals
7-day dissolving fish feeder tablet	5	10	5	5	1	26
Put a whole lot of fish food in at the beginning of the week	10	10	1	5	0	26
Hire a person to feed the goldfish	10	10	4	2	3	29
FishFeeder Timer Device (an original device using a timer and gravity feed of premeasured food)	9	5	9	5	2	30

Another system uses the following rating scale:

+ = good or better than other alternatives or positive influence
0 = fair or at par with other alternatives or neutral influence
• = poor or worse than other alternatives or negative influence

Try This: What are the Criteria?

Brainstorm design criteria for the problems listed in **Table 3**.

Table 3

Problem	Design Criteria
A device or process is needed that will make polluted water safe to drink.	?
A device is needed to keep a can of soda cold from breakfast until noon.	?
An optical device is needed that would allow a person to observe hatching birds in their nest.	?
A process or device is needed for watering hanging baskets of flowers.	?

3D Planning a Prototype

Once you know what alternative you want to design and build, it's time to plan a prototype. A prototype is an original model that exhibits the essential features of the solution you chose. It can be a device or it can be a method that demonstrates a new or an improved process.

- **Start by making sketches and detailed drawings of your prototype.** Initially, these may be hand-drawn sketches with approximate dimensions, but you may find it worthwhile to produce a prototype drawing that shows specific measurements and working parts. (Refer to 6C "Scientific & Technical Drawing" for more details.)

  You may want to use computer-aided design programs to help you in the design process.
- **Make a list of the materials and tools you will need to build your prototype.** Examine your drawings carefully to ensure that your list is as complete as possible.
- **Write out the instructions you will follow to build the prototype.** Think this through carefully. As you set down each step in the building process, you may think of additional materials and tools to add to your list. You may also uncover potential difficulties or problems that require you to make design changes. This is not a design failure! Regular evaluation and modification of your design is an important part of the process.
- **Have others review your plan.** A fresh pair of eyes will often discover small flaws or visualize slight improvements that will enhance your design. In your Science & Technology class, you will often be asked to have your teacher approve your plan before you begin building.

Try This **Planning**

Choose one of the problems in **Table 3** in section 3C and plan a prototype.

3E Building a Prototype

Before you begin construction, review the safety precautions for using the tools you'll need. Be sure that you have access to all the necessary materials and tools. If you are gathering these in advance, consider how you will store them for use throughout the project.

General Building Tips

- Read through your building instructions before you begin.
- If you need to, review specific tools (5D "Using Technology Equipment") or specific fabrication techniques (5E "Fabrication Techniques").
- Measure and cut materials to prepare for assembly. (Remember: Measure twice, cut once.)
- Temporarily join parts to test and evaluate their effectiveness. You may wish to use elastics, Plasticine, tape, string, and plastic tubing to assist you.
- Once you are satisfied with the components and their ability to operate, then you are ready to join the components permanently. It is important to consider methods of assembly that allow the parts to be added in such a way that they can be moved slightly, if needed.

Try This Deconstruction

One way of learning how to build things is to take things apart (**Figure 1**). Take apart one of the things listed below and jot down anything you learn about how it is put together.

- a flashlight
- a pen
- a chocolate box
- a bicycle wheel

Figure 1

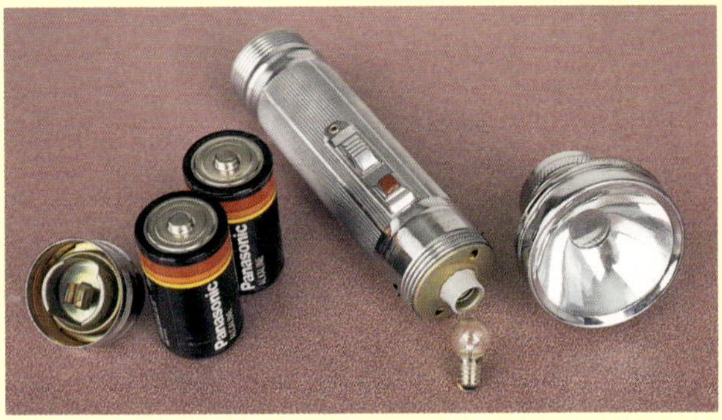

Consider the construction materials, how things are held together, the shape of the various parts, what tools would have been required, etc.

Do not take apart small appliances or electronics. You may receive an electric shock.

3F Testing and Evaluating a Prototype

What to Test

It is important to find out how well your prototype works. You need to develop a good test that will answer three basic questions:

1 Does the prototype solve the problem?
2. Does the prototype satisfy the design criteria?
3. Are there any unanticipated problems?

Does the Prototype Solve the Problem?

Let's assume you have built an automatic fish-feeder to solve the following problem:

A method or device is necessary to feed ten goldfish 0.5g of food, two times a day, for a period of one week.

To determine whether or not your prototype solved the problem, your test-period would have to be at least one-week long, you would have to record how many times a day food actually made it into the tank, and you would have to record the amount of food that was deposited at each feed.

Does the Prototype Satisfy the Design Criteria?

Let's assume that you identified and rated the importance of the design criteria in **Table 1**:

Table 1

time:	maximum importance	(rated between 0-10)
cost:	maximum importance	(rated between 0-10)
reliability:	maximum importance	(rated between 0-10)
ease of use:	medium importance	(rated between 0-5)
appearance:	low importance	(rated between 0-3)

To determine whether the prototype meets the design criteria, your test must measure against them. In some cases, the measure will be quantitative (e.g., time and cost). In other cases, your measure will be qualitative (e.g., appearance). In any case, your evaluation must be carefully recorded.

Are There Any Unanticipated Problems?

Your test may also show you weaknesses that you hadn't considered. For instance, does the position of the fish-feeder on the tank affect how much light is reaching the water? Do you need to adjust the position? Do you need to reconsider the materials you have used?

Testing Tips

The following tips will help you design and conduct your tests:

- Decide how long the test should last.
- Decide how many times the test should be done.
- Consider whether the test needs to be done in different circumstances or conditions.
- Rate each of the design criteria.
- Make and record all quantitative measurements.
- Record all qualitative observations.
- Consider potential problems that may not be related to the design criteria.
- Have other people make use of the prototype and get their feedback.
- Keep a log of your test information.

Modifying Your Design

Your test will probably give you ideas about how your design can be improved. This doesn't mean that you are a bad designer. The design process is a repeating cycle, not a step-by-step path to the perfect solution. You will need to go through as many repetitions of the problem-solving cycle as needed in order to come up with the best working solution. Once you have tested your prototype, it is time to re-address the problem statement, redefine your design criteria, and work on improving your device or process.

Try This Building the Best Papercopter

A babysitter is told that his young charge can unexpectedly become very unhappy. When this happens, the only thing that will console him is watching things, like paper airplanes, fly. The babysitter can only make one kind of paper airplane, and it doesn't fly very well. The child is unimpressed. So, the babysitter's problem is to design a device that would fly long enough to amuse the child (about 10 s).

The design criteria include ease of construction, speed of construction, reliability, and accessibility of materials.

1. Follow the babysitter's plans (**Figure 1**) and build his prototype: the papercopter.
2. Now test the papercopter.
   - Make a list of things to measure and observe.
   - Conduct the test.
   - Summarize what needs to be improved.
   - List modifications to try.
   - Modify the papercopter until it solves the problem.

Figure 1

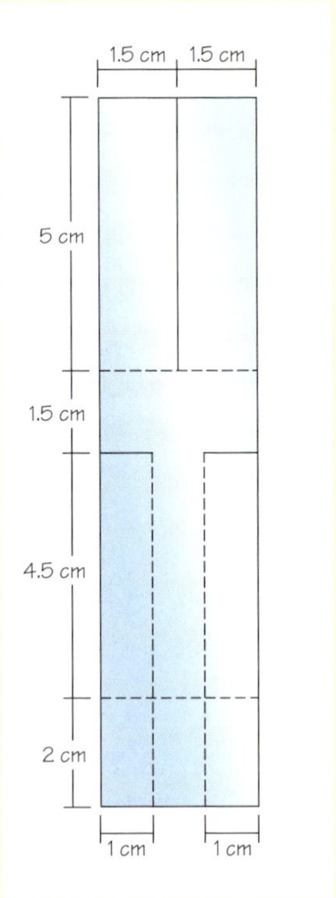

1. Cut a piece of writing paper 3 cm x 13 cm.
2. Cut along the two middle horizontal (solid) lines, and fold in.
3. Fold the bottom flap along the bottom dotted line.
4. Cut the solid vertical line at the top to create two flaps.
5. Attach a paperclip to the bottom.
6. Drop and watch the copter fly.

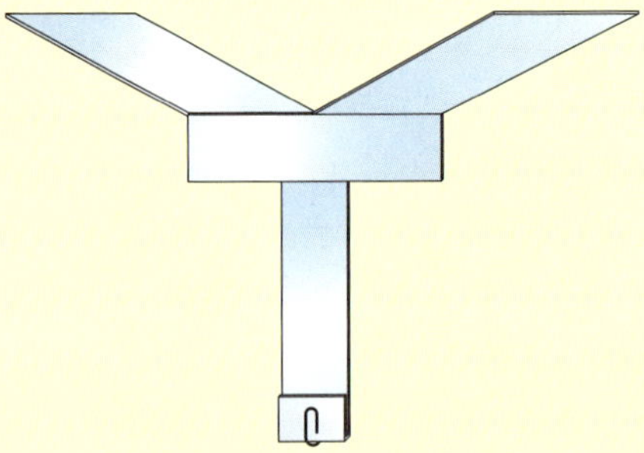

3G Patents, Trademarks, and Copyrights

What would you do if you invented an innovative product or process? Imagine you invented a "Landsurfer" and wanted to develop, sell, and market your invention. In our society, the law protects your intellectual property for a certain period of time. Anyone designing new products or processes should know the differences between and the uses of patents, trademarks, and copyrights.

Patents

A patent gives an inventor the exclusive rights to make, use, and/or sell a certain product, device, or process for a period of 17 years. The inventor must apply for the patent through the Canadian Patent Office. It is usually necessary to contact a Patent Attorney or Agent in order to search for other existing patents on your product and then to file the legal application. The inventor must prove that the invention is original, beneficial to society, and "not obvious" (i.e., something that is in common use, but has never previously been patented). The patent itself includes a number, title, and a short description. You may have noticed the words "Patent Pending" on a product or device. Because the patent process can be a lengthy one, inventors can let others know that they have applied for a patent by printing these words on the device.

Trademarks

A trademark is any name, symbol, picture, or design that an inventor or company may use to identify their products or affairs (**Figure 1**). Trademarks are often indicated by the following symbol: ™. Some well known and much used trademarks often become the actual name of a product and thus lose their legal status as trademarks. For example, aspirin and escalator were originally trademarks and have since become generic product names.

Figure 1

Nelson
Thomson Learning™

Copyright

A copyright protects the work of artists, composers, and authors from reproduction and/or distribution by others without permission (**Figure 2**).

Figure 2

Copyright © 2000 Nelson, a division of Thomson Learning. Thomson Learning is a trademark used herein under license.

ALL RIGHTS RESERVED. No part of this work covered by the copyright hereon may be reproduced, transcribed, or used in any form or by any means—graphic, electronic, or mechanical, including photocopying, recording, taping, Web distribution or information storage and retrieval systems—without the permission of the publisher.

Try This — Find Examples

- List 10 inventions, devices, or products that would most likely already have a patent. You may want to visit the Canadian Patent Office's web site to see some examples of current patents.
- List 10 trademarks with which you are familiar.
- List 10 things that might be copyrighted.

4A Research Skills

Sources of Information

There is an incredible amount of scientific and technological information available to you. Before you can use it, however, you have to know how to gather it efficiently. First, you need to know where to find information.

Look at the list of information resources in **Table 1**. Some resources are more useful than others in certain circumstances. Pick one or two resources that you seldom use and think of how you might use them in your Science & Technology class.

Before you start any research, brainstorm a list of the best resources for your purpose. Rank the list, starting with the most useful resource. Always use a variety of resources.

General Research Tips

Once you've decided where to look for information, you need to find the best information in the least amount of time. Using these tips will help:

- Clearly define the topic or the question you will be researching. Ask yourself: "Do I clearly understand what I'm looking for?"
- List the most important words associated with your research, and use them to search for related topics.
- Ask yourself: "Do I understand what this resource is telling me?"
- Check when the resource was published. Is it up-to-date?
- Keep organized notes while doing your research.
- Keep a complete list of the resources you used, so you can make a bibliography when writing your report. (See 8A "Writing a Report" for proper referencing formats.)
- Review your notes. After your research, you may want to alter your original problem or hypothesis.

Table 1 Information Resources

Information Consultants (people who can help you locate and interpret information)	
teachers	business people
nurses	scientists
public servants	librarians
volunteers	politicians
lawyers	farmers
parents	doctors
members of the media	senior citizens
veterinarians	

Reference Materials (sources of packaged information)	
encyclopedias	bibliographies
magazines/journals	newspapers
videotapes	slides
data bases	almanacs
yearbooks	dictionaries
textbooks	maps
filmstrips	charts
biographies	films
pamphlets	records
television	radio

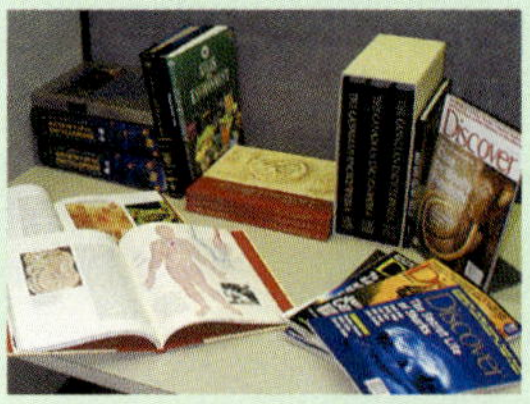

Places (sources beyond the walls of your school)		
public libraries	shopping malls	art galleries
parks	colleges	museums
universities	government offices	hospitals
historic sites	zoos	volunteer agencies
research laboratories		
businesses		
farms		

Electronic Sources
world wide web (www)
CD-ROMs
on-line search engines
on-line periodicals
computer programs

Some Specific Research Tips

Using a Library

School libraries use an organizational system, known as the Dewey Decimal Classification System, which organizes books into the major subject areas shown in **Table 2**.

Table 2 **Dewey Decimal Classification System**

Catalogue #	Subject Area	Catalogue #	Subject Area
000	Generalities	500	Natural Science & Mathematics
100	Philosophy & Psychology	600	Technology (Applied Sciences)
200	Religion	700	The Arts
300	Social Science	800	Literature
400	Language	900	Geography/History

You can usually search a library database by title, author, and/or subject area. If you have access to a computer, many library databases can be accessed online.

Always ask the librarian for help if you are having trouble finding the information you need.

Using On-line Sources

The Internet is a vast network of information that is continually growing. Many Internet navigation programs make use of several search engines. Search engines are on-line tools that find web sites (or hits) based on key words you enter into the search. When researching on-line, consider these additional research tips.

- If you have a specific web site address, go to the direct source first.
- If you need to use a search engine, find two good search engines, bookmark them, and use them. While the "Search" buttons in navigation programs are useful, they will not always give you the best search engine for your purposes.
- Learn about the features of the search engine. The better search engines make use of Boolean logic, operations that are used to combine key words when searching the Internet. In these cases, adding the word "AND" can let you narrow your search by combining two key words. Adding the word "OR" can let you expand your search by joining together two key words. Lastly, adding the word "NOT" enables you to disregard a key word. Be sure to refine your search as you continue. For example, entering the word "music" may give you 1 678 243 hits but adding "AND guitar" will reduce your search results to 18 860 hits. Including "NOT classical" will further reduce your number of hits to 2340.
- Familiarize yourself with the advanced search option within each search engine.
- Look over the first few pages of hits before you start exploring each web site. Be patient and look for the sites that seem most relevant to your topic.
- Bookmark all useful sites so you can access them readily. Put these in a specific folder if your browser allows you to do this.

Try This **Research This!**

Use the General Research Tips to research one of the following topics:

- the discovery of insulin
- eco-tourism in Costa Rica
- the invention of the telescope
- the most recent seismic activity on Earth
- the tides in the Bay of Fundy
- how pneumatics and hydraulics can be applied to everyday life
- how valves and pumps work
- the physical properties of plastics
- the manufacturing process of the product of your choice

Be prepared to communicate the information you find, bibliographic references (see 8A "Writing a Report"), how reliable your information is, and your research process.

4B Interviewing and Survey Skills

When to Use Surveys and Interviews

Surveys and interviews give you access to a very important information resource: people. If you need specialized information, interviewing experts (for example, engineers or researchers) is useful. If you want to find out if your design idea fills a common human need, a survey is an excellent tool. Surveys and interviews are both useful for learning about people's reactions to controversial scientific and technological issues.

Generally, surveys are better if you need specific information from a lot of people. Interviews are better if you need more detailed information from fewer people.

Effective Interviews

Use the following suggestions to help you decide who you are going to interview, and how to prepare and conduct the interview.

- ***Research your topic.*** Clarify what you need to learn through your interview(s).
- ***Choose your interview candidate(s).*** For instance, if you are researching water quality issues in your community, you might want to interview the municipal waterworks engineer, the health inspector, a member of a local environmental group, and a journalist who has written articles on the topic. Make contact with these people: explain your purpose clearly and arrange a convenient time and place.
- ***Prepare your interview questions in advance.*** Try to avoid simple yes/no questions—questions that require explanations will provide more information.
- ***Conduct your interview in a comfortable, quiet place.***
- ***Be prepared to ask questions that are not on your list.*** You want to make sure all of your questions are answered, but some answers will lead to other important questions. For example, if your interviewee said "The water quality is well within set specifications and much better than it was after that incident ten years ago," you may want to ask about the "incident." A good interviewer learns when to let the interview follow its own course and when to bring it back to the prepared questions.
- ***Choose an appropriate way of recording the information.*** You may want to take notes or use a tape recorder. Be sure to ask if it is okay to tape record the interview. If you take notes, be sure that this does not distract you or the person you are interviewing. If you do not understand an answer, be sure to politely ask for clarification.
- ***Be aware of your own biases as well as the biases of the person you are interviewing.*** Remember every person has opinions that have been shaped by personal experience—these are called biases. These will influence the information you get and how you respond to it.
- ***If you quote people, use their exact words and make sure you have their permission.***
- ***Be respectful of the person you are interviewing.*** End your interview on time and express your thanks in person and later with a thank-you note.

Effective Surveys

Planning the Survey

An effective survey requires careful planning. Try to think it through from beginning to end, using the following questions to help you:

- What exactly are we trying to find out?
- Who should we be surveying?
- How many people should we survey?
- Should we ask the survey questions in person and fill in the forms or send the surveys out to be filled in by the people being surveyed?
- How will we use the information? How will we want to present the data?

Answering these questions will affect how you design your survey questions.

Writing the Survey Questionnaire

Writing a good questionnaire requires skill and practice. Consider the following:

- ***Design your questionnaire so that it can be completed easily and quickly.*** The questions should be precise and concise, and the instructions should be easy to understand. Start with questions that can be answered by checking off options or giving a numbered rating, and leave the more open-ended questions to the end.
- ***Design your questionnaire so that it is easy to collate and present the results.*** It should be possible to present your results in meaningful charts and graphs. Using an appropriate software program can help you design your questionnaire and collate your results.
- ***Begin your questionnaire with questions about the respondent.*** Be sure that these questions are relevant to the topic you are studying.
- ***Write clear questions that are easy to answer.*** In order to quantify your results easily, you may want to use the Likert scale, developed by Renis Likert (1932). Start by writing positive or negative statements that relate to the issue being explored. Be sure to minimize any bias in your statements. For example, "CFCs should be banned in Canada" might be better than "Harmful CFCs should be banned in Canada."

  Then, attach a five-point scale to rate public reaction to this statement. For example:

  Circle the number that best describes your reaction to the following statement:

  CFCs should be banned in Canada.

  1 2 3 4 5

  where

  1 = strongly agree, 2 = agree,
  3 = undecided, 4 = disagree, and
  5 = strongly disagree

- ***Be courteous.*** If you are sending your survey in the mail, you may want to write a covering letter that explains your research and thanks the respondent in advance. If you are conducting the survey in person, use common courtesy.

Testing the Questionnaire

It is often a good idea to try your questionnaire on three or four people before sending it out to everyone on your list. You may need to modify the questionnaire after you have considered the following questions:

- How well did the respondents understand the questions?
- Did they give the kind of information you need?
- How long did it take to complete?

Try This Interview or Survey?

Consider the following research questions. Would you use interviews, surveys, or both to find the answers?

- What kind of training do you need to be a water resource engineer?
- How satisfied are e-mail users with their Internet service provider?
- Should we ban the use of CFCs?
- How much of Ontario's land should be set aside as protected park land?
- Should safety ever be compromised to lower costs?
- How well is the local recycling program working?
- What technology do most people feel has the greatest impact on their daily lives?

4C Critical Thinking

Science & Technology in the News

Understanding and evaluating the results of your research is extremely important. When you do research, you can find information using the Internet, textbooks, magazines, chat lines, television, radio, and many other forms of communication. Is all of this information reliable? How do we know what to believe and what not to believe?

Every day you see and hear extraordinary claims about objects and events (**Figure 1**). Often scientific or technological evidence is used to convince us that these claims are true. This method of reporting is sometimes used to catch your attention or to make you buy something. Even serious stories on scientific research or technological innovation can be difficult to interpret, especially when they are reported in a way that makes the work sound important, official, and somewhat mysterious.

Figure 1

There are scientific and technological stories in the news every day.

To analyze information, you have to use your mind effectively and critically. When you encounter a report that uses scientific or technological evidence, analyze the report carefully. You might ask yourself, for instance:

- Was the investigation conducted by a reputable laboratory or firm?
- How many times was the design or inquiry process repeated?
- What are the possible sources of error? Are they addressed?
- What do other experts say about these claims?

Try This — Reading Critically

Analyze a scientific report or story from one of the following magazines by answering the questions listed above:

- *Scientific American*
- *Popular Mechanics*
- *Discover Magazine*
- *Eureka: The Canadian Invention and Innovation Newsletter*
- or a magazine of your choice

Using Critical Thinking Skills

A lot of research has been done to help people develop critical thinking skills. One very good, practical framework is known as PERCS (which comes from the word perquisite, meaning a benefit). It was founded out of Central Park East Secondary School in New York City. To prepare themselves for their world, the students at CPESS use a series of questions to help them think critically about information and arguments. These questions will help you build an educated and critical opinion about an issue.

The PERCS Checklist:

P = Perspective

- From whose viewpoint are we seeing or reading or hearing?
- From what angle or perspective?

E = Evidence

- How do we know what we know?
- What's the evidence and how reliable is it?

R = Relevance

- So what?
- What does it matter?
- What does it all mean?
- Who cares?

C = Connections

- How are things, events, or people connected to each other?
- What is the cause and what is the effect?
- How do they "fit" together?

S = Supposition

- What if...?
- Could things be otherwise?
- What are or were the alternatives?
- Suppose things were different?

Try This Using PERCS

Try exercising your critical thinking skills. Choose any article from today's newspaper or from a popular magazine and analyze it using the PERCS checklist.

5A Using the Microscope

Because cells are small, you must make them appear larger than they really are in order to see and study them. To view cells closely, you will use a compound light microscope (**Figure 1**). It employs two lenses and a light source to make the object appear larger. The object is magnified by a lens near your eye, the ocular lens (sometimes called the eyepiece), and again by a second lens, the objective lens, which is just above the object. The comparison of the actual size of the object with the size of its image is referred to as magnification.

The parts of the microscope are described in **Table 1**.

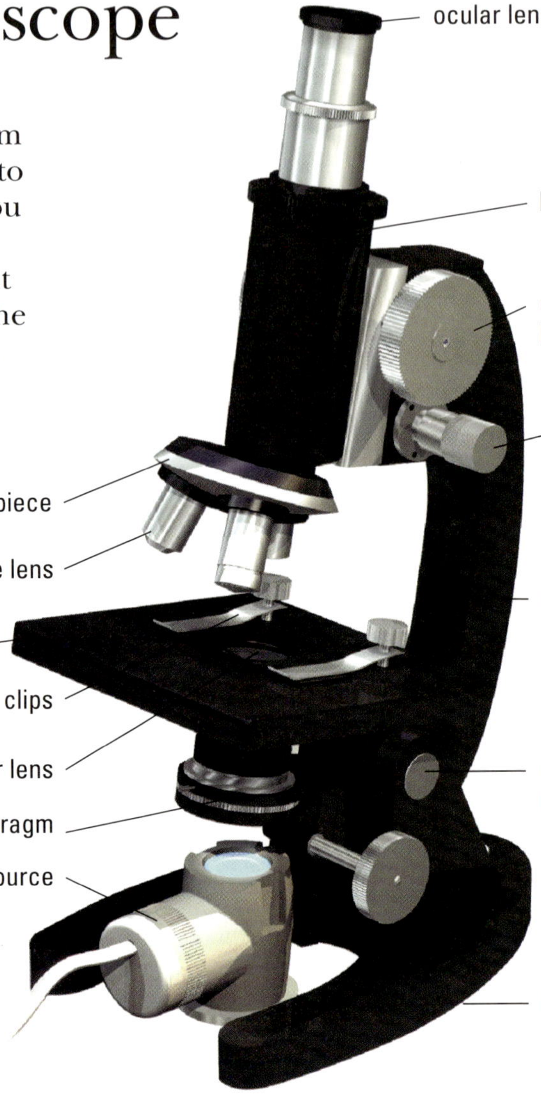

Figure 1 The microscope

Table 1 **Parts of the Microscope**

Structure	Function
body tube	Contains ocular lens, supports objective lenses.
clips	Found on the stage and used to hold the slide in position.
coarse-adjustment knob	Moves the body tube up or down so you can get the object or specimen into focus. It is used with the low-power objective lens only.
condenser lens	Directs light to the object or specimen.
diaphragm	Regulates the amount of light reaching the object being viewed.
fine-adjustment knob	Moves the tube to get the object or specimen into sharp focus. It is used with medium- and high-power magnification. The fine-adjustment knob is used only after the object or specimen has been located and focused under low-power magnification using the coarse adjustment.
objective lenses	Magnifies the object. Usually three complex lenses are located on the nosepiece immediately above the object or specimen. The smallest of these, the low-power objective lens, has the lowest magnification, usually four times (4X). The medium-power lens magnifies by 10X, and the long, high-power lens by 40X.
ocular lens	Magnifies the object, usually by 10X. Also known as the eyepiece, this is the part you look through to view the object.
revolving nosepiece	Rotates, allowing the objective lens to be changed. Each lens clicks into place.
stage	Supports the microscope slide. A central opening in the stage allows light to pass through the slide.

Care of the Microscope

The microscope is an important and expensive scientific instrument. It should always be used with care and patience. Here are some points to remember when using your microscope.

1. Always keep a microscope in an upright position.
2. When carrying a microscope, grasp its arm with one hand and support its base with the other (**Figure 2**).

Figure 2
How to carry a microscope

3. Once you have the microscope at your workstation, turn the coarse adjustment knob to lower the objective lens to its lowest point. Focus by using the knob to move the objective lens slowly upward to bring an object into focus. You must always remember to use both knobs when adjusting your microscope or you will strip the adjustment gears.
4. Microscope lenses are made of optical glass, which is soft and scratches easily. Use special lens paper to remove any dust or dirt.
5. When you complete an investigation using the microscope, follow these steps:
   - Rotate the nosepiece to the low-power objective lens.
   - Remove the slide and cover slip (if applicable).
   - Clean the slide and cover slip and return them to their appropriate location.
   - Return the microscope to the storage area.

Basic Microscope Skills

The skills outlined below are presented as sets of instructions. This will enable you to practise these skills before you are asked to use them in the investigations in *Nelson Science & Technology 7/8.*

Materials

- newspaper that contains lower-case letter "*f*" or similar small object
- scissors
- microscope slide
- cover slip
- dropper
- water
- compound microscope
- thread
- compass or petri dish
- pencil
- transparent ruler

Preparing a Dry Mount

This method of preparing a microscope slide is called a dry mount, because no water is used. A dry mount can be used for any specimen that won't dry out while you are examining it.

1. Find a small, flat object, such as a lower-case letter "*f*" cut from a newspaper.
2. Place the object in the centre of a microscope slide.
3. Hold a cover slip between your thumb and forefinger. Place the edge of the cover slip to one side of the object. Gently lower the clover slip onto the slide so that it covers the object.

Step 3

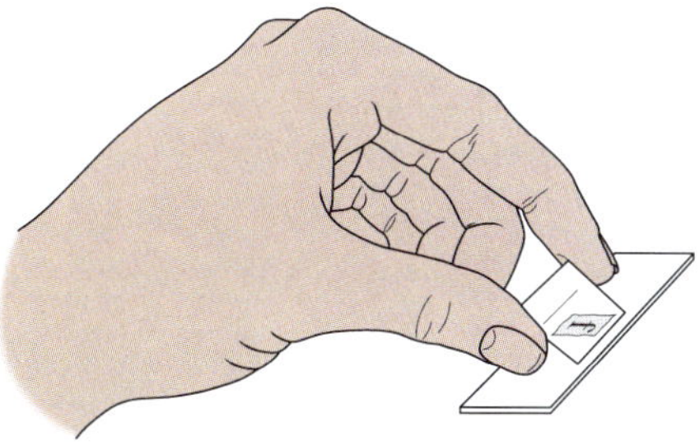

Preparing a Wet Mount

This method of preparing a microscope slide is called a wet mount, because water is used.

A wet mount is used for specimens that can dry out while you are examining them. For instance, when you use a specimen from a living thing, like an onion, the specimen must be very thin. This means that it can lose moisture rapidly: it needs to be kept moist or it will begin to shrivel and become more and more difficult to examine.

1. Find a small, thin, flat object.
2. Place the object in the centre of a microscope slide.
3. Place two drops of water on the object.

Step 3

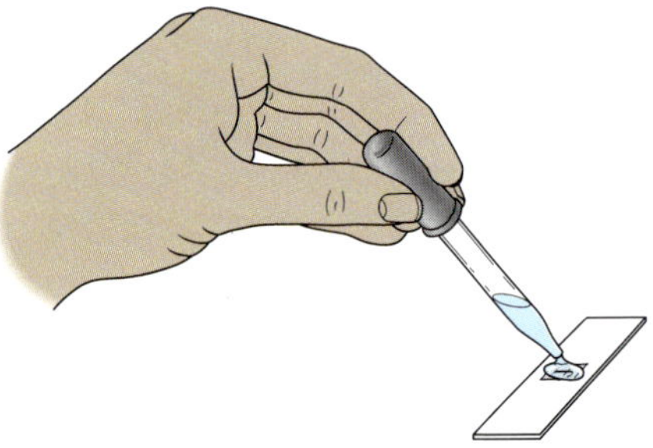

4. Holding the cover slip with your thumb and forefinger, touch the edge of the surface of the slide at a 45º angle. Gently lower the cover slip, allowing the air to escape.

Step 4

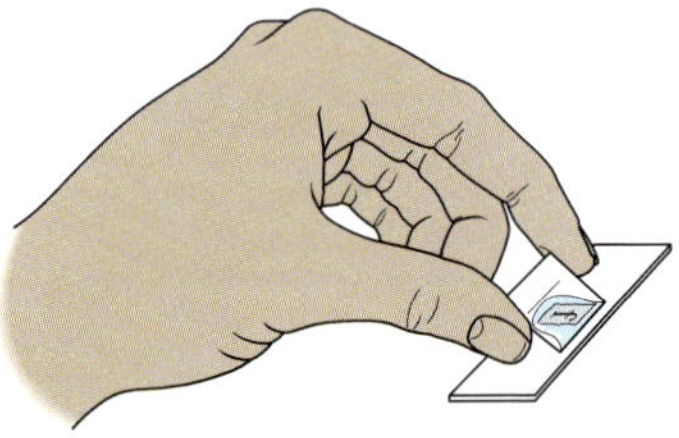

Positioning Objects Under the Microscope

1. Make sure the low-power objective lens is in place on your microscope. Then put either the dry or wet mount slide in the centre of the microscope stage. Use the stage clips to hold the slide in position. Turn on the light source.

Step 1

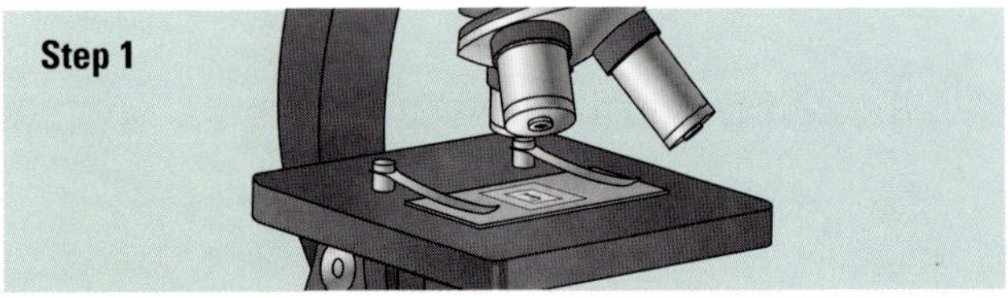

2. View the microscope stage from the side. Using the coarse-adjustment knob, bring the low-power objective lens and the object as close as possible to one another. Do not allow the lens to touch the cover slip.

Step 2

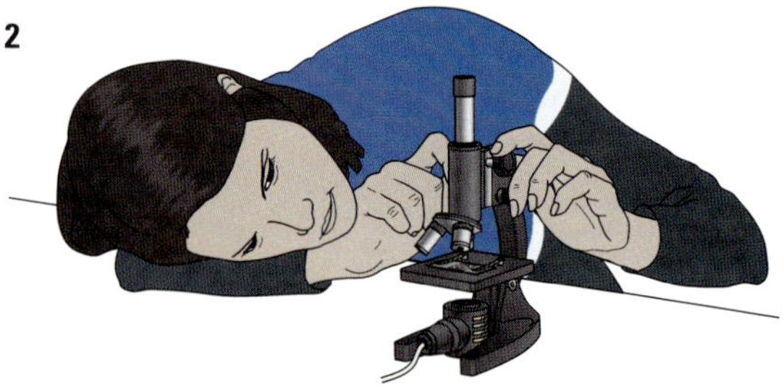

3. View the object through the eyepiece. Slowly move the coarse-adjustment knob so the objective lens moves away from the slide, to bring the image into focus. Note that the object is facing the "wrong" way and is upside down.
4. Using a compass or a petri dish, draw a circle in your notebook to represent the area you are looking at through the microscope. This area is called the field of view` Look through the microscope and draw what you see. Make the object fill the same amount of area in your diagram as it does in the microscope.
5. While you are looking through the microscope, slowly move the slide away from your body. Note that the object appears to move toward you. Now move the slide to the left. Note that the object appears to move to the right.
6. Rotate the nosepiece to the medium-power objective lens. Use the fine-adjustment knob to bring the object into focus. Note that the object becomes larger.

Never use the coarse-adjustment knob with the medium or high-power objective lenses.

7. Adjust the object so that it is directly in the centre of the field of view. Rotate the nosepiece to the high-power objective lens. Use the fine-adjustment knob to focus the image. Note that you see less of the object than you did under medium-power magnification. Also note that the object seems closer to you.

Investigating Depth of Field

The depth of field is the amount of an image that is in sharp focus when it is viewed under a microscope.

1. Cut two pieces of thread of different colours.
2. Make a temporary dry mount by placing one thread over the other in the form of an X in the centre of a microscope slide. Cover the threads with a cover slip.

Step 2

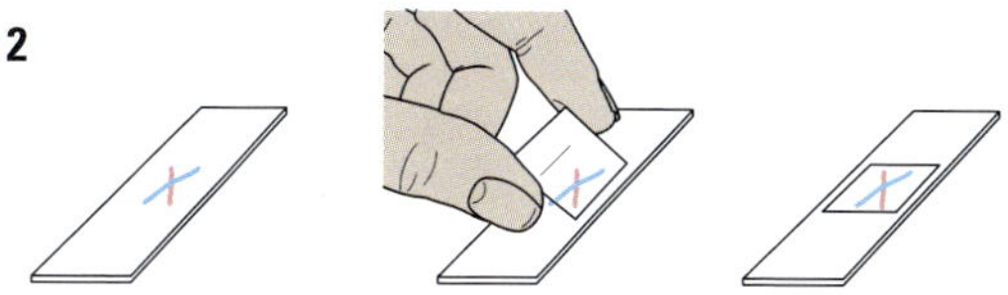

3. Place the slide on the microscope stage and turn on the light.
4. Position the low-power objective lens close to, but not touching, the slide.
5. View the crossed threads through the ocular lens. Slowly rotate the coarse-adjustment knob until the threads come into focus.
6. Rotate the nosepiece to the medium-power objective lens. Focus on the upper thread by using the fine-adjustment knob. You will probably notice that you cannot focus on the lower thread at the same time. The depth of the object that is in focus at any one time represents the depth of field.
7. Repeat step 6 for the high-power objective lens. The stronger the magnification, the shallower the depth of field.

Determining the Field of View

The field of view is the circle of light seen through the microscope. It is the area of the slide that you can observe.

1. With the low-power objective lens in place, put a transparent ruler on the stage. Position the millimetre marks on the ruler immediately below the objective lens.
2. Using the coarse-adjustment knob, focus on the marks on the ruler.
3. Move the ruler so that one of the millimetre markings is just at the edge of the field of view. Note the diameter of the field of view in millimetres, under the low-power objective lens.

Step 3

4. Using the same procedure, measure the field of view for the medium-power objective lens.
5. Most high-power lenses provide a field of view that is less than one millimetre in diameter, so it cannot be measured with a ruler. The following steps can be followed to calculate the field of view of the high-power lens.

   Calculate the ratio of the magnification of the high-power objective lens to that of the low-power objective lens.

$$\text{Ratio} = \frac{\text{magnification of high-power lens}}{\text{magnification of low-power lens}}$$

   Use the ratio to determine the field of diameter (diameter of the field of view) under high-power magnification.

$$\text{Field diameter (high power)} = \frac{\text{field diameter (low power)}}{\text{ratio}}$$

Estimating Size

1. Measure the field of view, in millimetres, as shown above.
2. Remove the ruler and replace it with the object under investigation.

Step 2

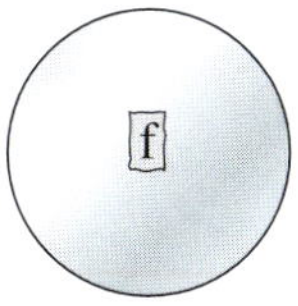

3. Estimate the number of times the object could fit across the field of view.
4. Calculate the width of the object:

$$\text{width of object} = \frac{\text{width of field of view}}{\text{number of objects across field}}$$

Remember to include units.

5B Working with Scales and Balances

Mass is an important quantitative property that is frequently measured: it is the amount of matter a substance has. A balance will give you the same measurement of mass no matter where you are, because it compares the object to another object of fixed mass acting under the same gravitational force. The electronic balance, the platform balance, and the triple beam balance are commonly used in Science & Technology classrooms.

Weight is different from mass. Weight is the measure of the force of gravity working on an object. An object will weigh different amounts in a valley, on the top of Mount Everest, and on the Moon—but it will have the same mass. Spring scales are often used to weigh things in Science & Technology classrooms.

The Electronic Balance

Electronic balances are easy to use, but they are also usually expensive and sensitive pieces of equipment (**Figure 1**). To operate an electronic balance, you simply place the sample on the platform and read the measurement on the digital display. If you want to measure the mass of a substance without including the mass of the container it is in, you can place the container on the platform and "re-zero" or "TARE" the balance. This resets the scale to zero. You can then measure the mass of the substance that you add.

Figure 1
An electronic balance

The Platform Balance

A platform balance (or equal arm balance) operates on the same principles as old scales used in commerce and, figuratively, in justice. Look at **Figure 2**. Can you see why it is called a "balance"?

Figure 2
A platform or equal arm balance

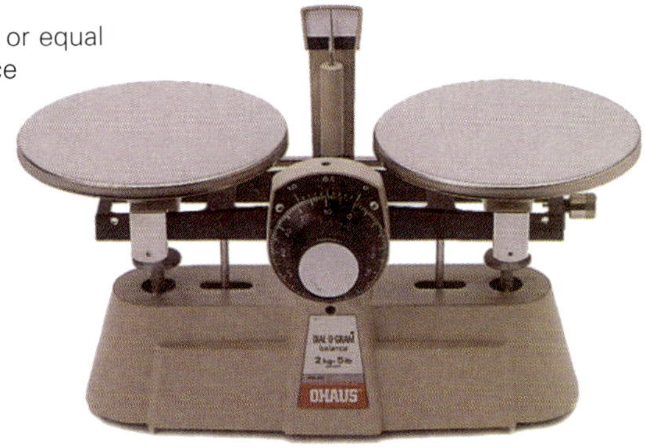

To use a platform balance, you balance your sample with known masses. Be sure that your balance is "balanced" before you use it—you may have to adjust the adjustment weights. Add masses in small amounts to come as close as possible to a balanced state. Then adjust the rider to make the final balance.

The Triple Beam Balance

The triple beam balance uses the same principle as the platform balance, but it also uses the principle of leverage (**Figure 3**). You measure

Figure 3
A triple beam balance

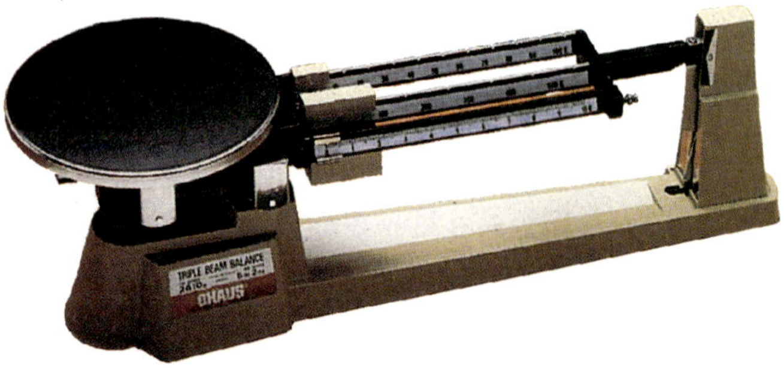

the mass of a substance by systematically adjusting the three rider beams (starting with the largest scale and ending with the smallest scale) and combining the three readings (**Figure 4**). Again, you may have to initially "zero" the scale by adjusting the adjustment weights.

Figure 4

Measuring the mass of a beaker

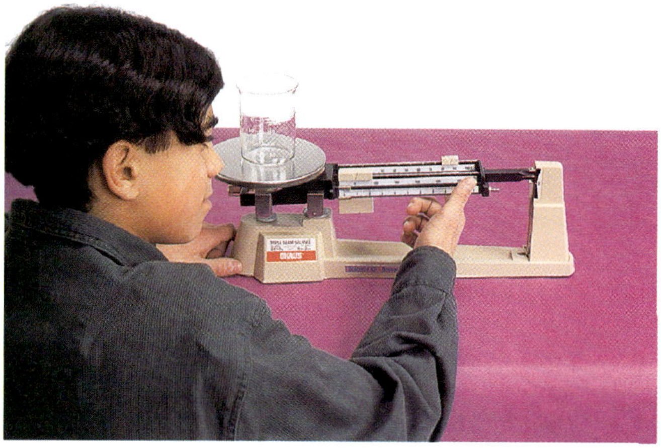

Try This: Measure Mass and Weight

Using one of the types of balances listed above, measure the following:

- the mass of a key
- the mass of the water in a glass (you may have to think about this if you are using either a platform balance or a triple beam balance)
- the mass of a spoon of sugar

How might you measure the weight of an object in water?

The Spring Scale

The spring scale is used to measure weight (**Figure 5**). In other words, a spring scale compares the force exerted by gravity (and any other forces that are acting—such as the buoyant force) to the counterforce applied by the spring and tells you where they balance, so the measurement varies with the force of gravity. It is particularly useful for items that do not fit on platforms or are different in shape. Spring scales come in various sizes. To use the scales, place the item to be measured on the hook of the spring scale and read the weight on the scale in the clear section of the tube.

Figure 5

Spring scales

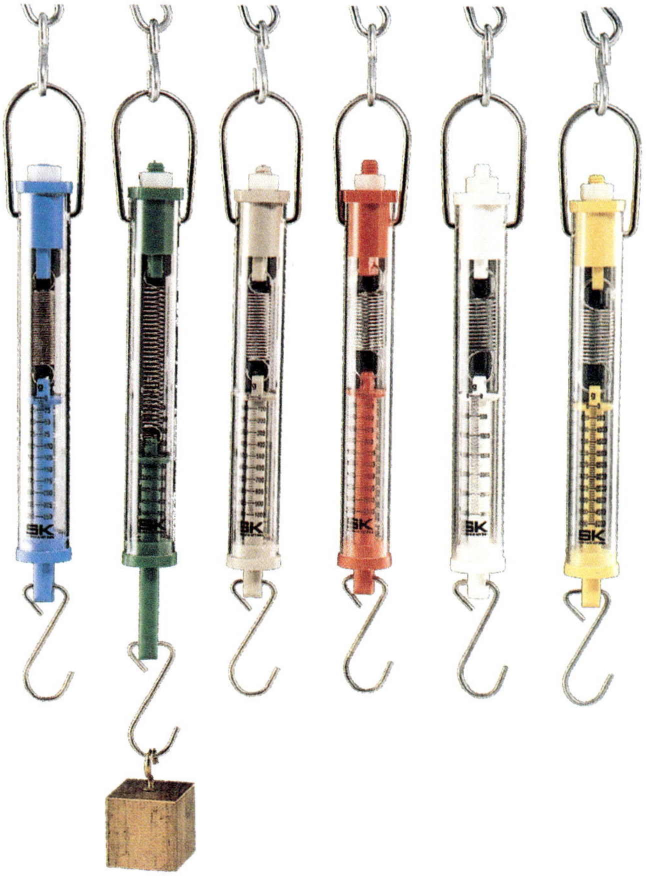

5C Using Other Scientific Equipment

You can do many investigations using everyday materials and equipment. In your Science & Technology classroom, there are other pieces of equipment. Some of these are illustrated below.

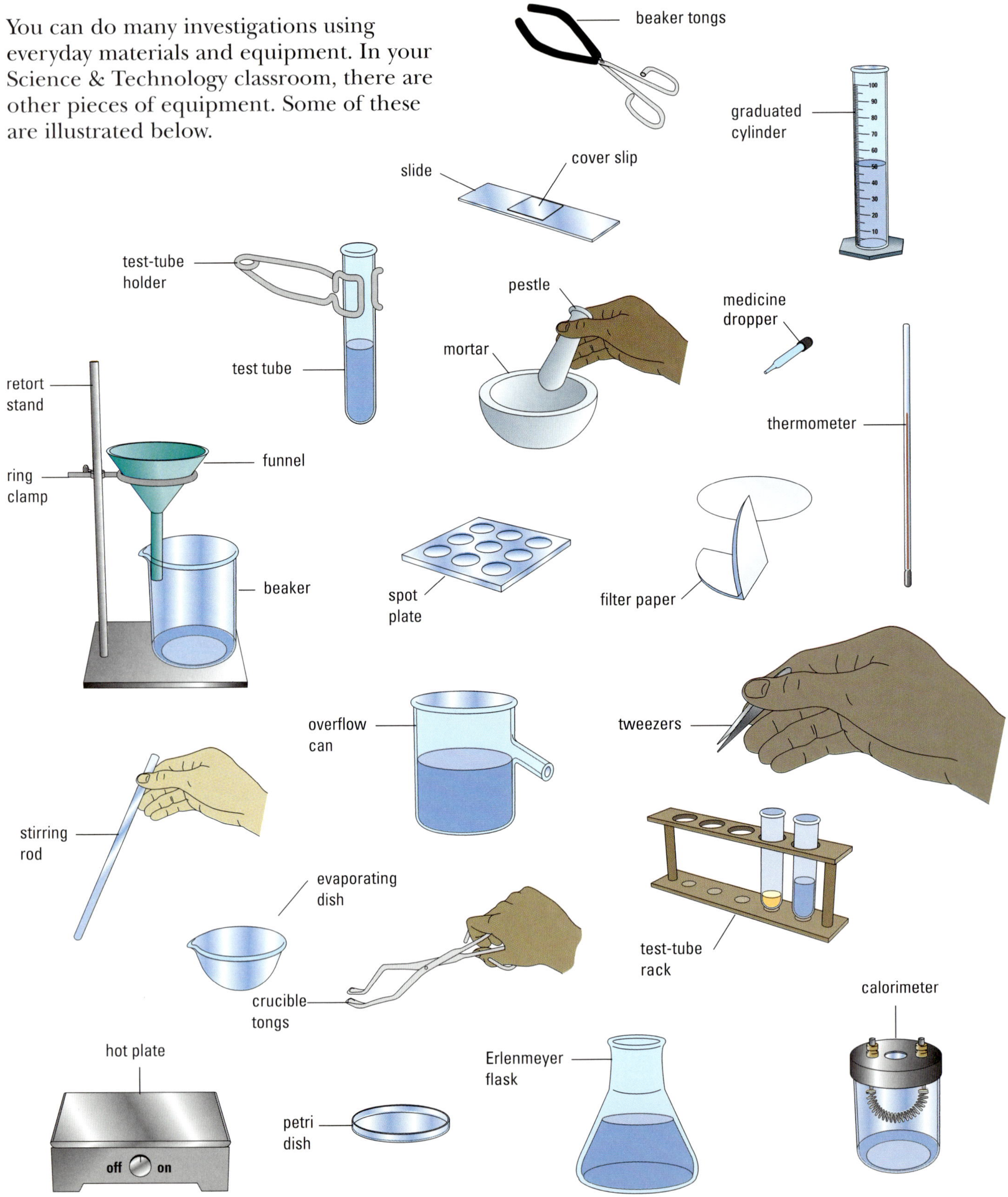

5D Using Technology Equipment

As you go through the design process, you will use various tools and materials to construct your prototypes. Some of these are illustrated below. (Refer to 1 "Safety in Science & Technology," to review the safe use of saws, drills, paper drills, utility knives, snips, hammers, and screwdrivers.)

Tools

Sawing and Cutting

hacksaw

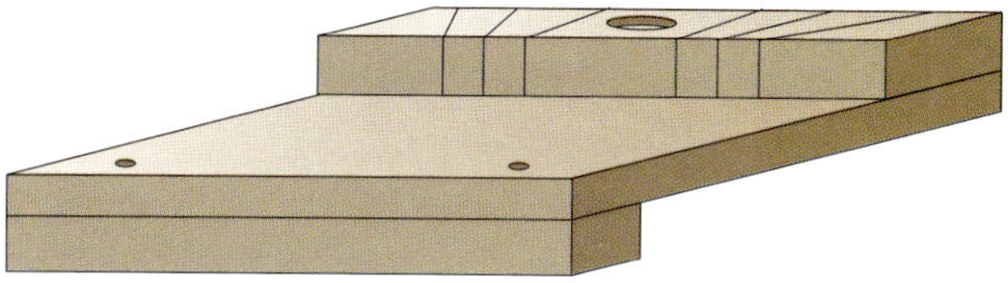

bench hook with mitre box

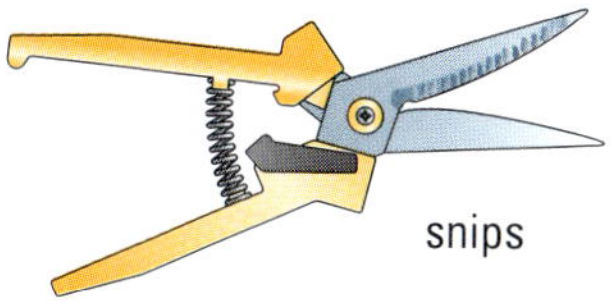

snips

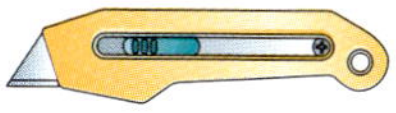

utility knife

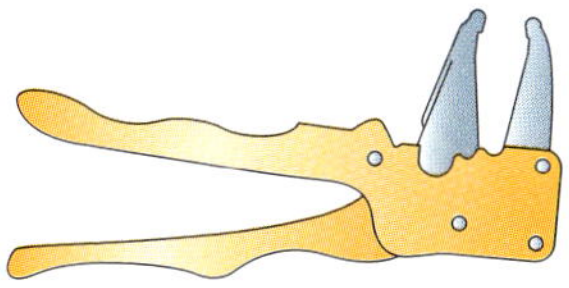

wire strippers

Joining

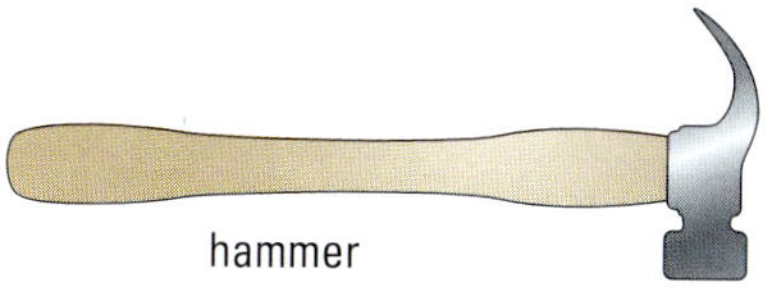

hammer

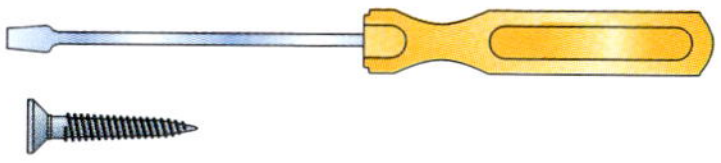

screwdriver and screw

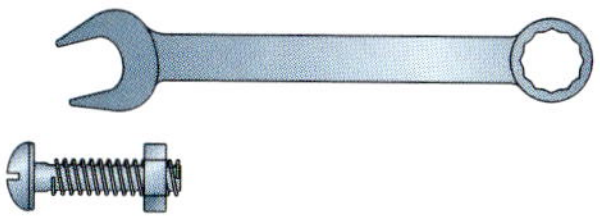

wrench, nut and bolt

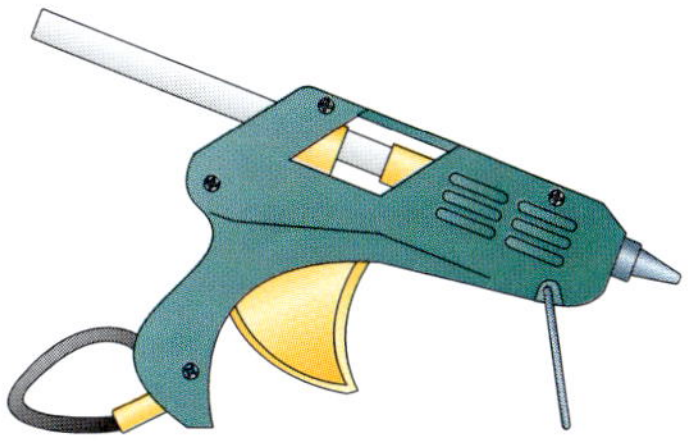

glue and glue gun

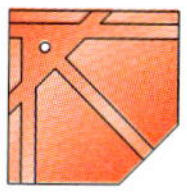

wood jointer

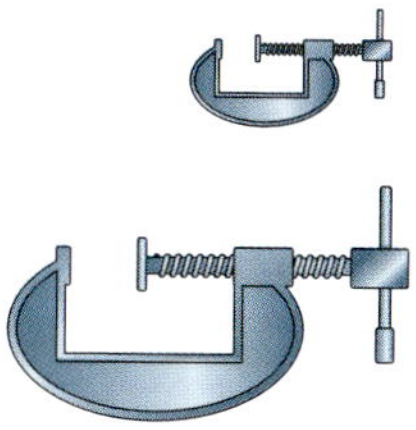

"C" clamps

Shaping

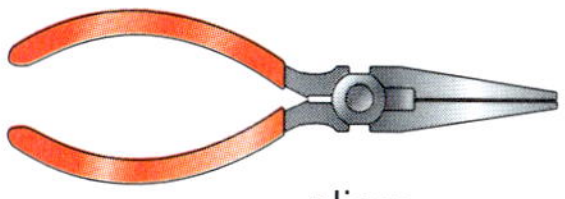

pliers

Measuring

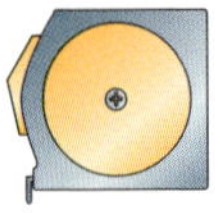

measuring tape

metal safety ruler

Making Holes

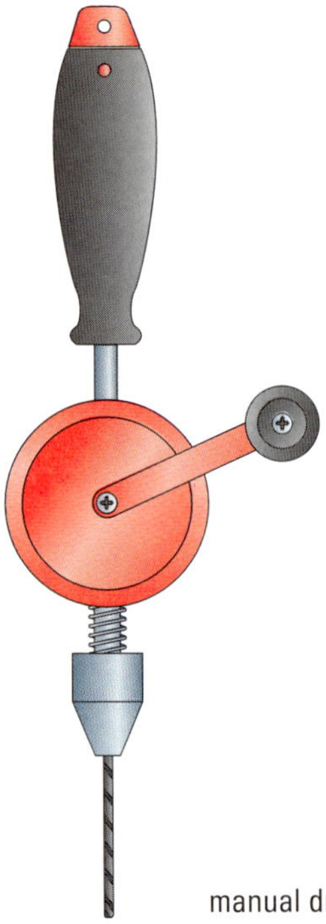

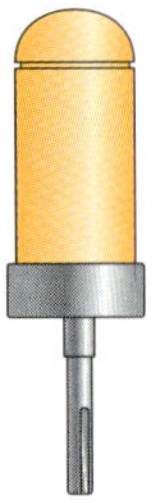

paper drill

manual drill and drill jig

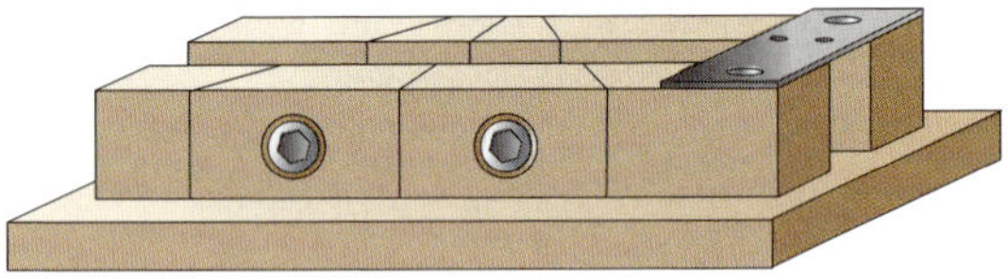

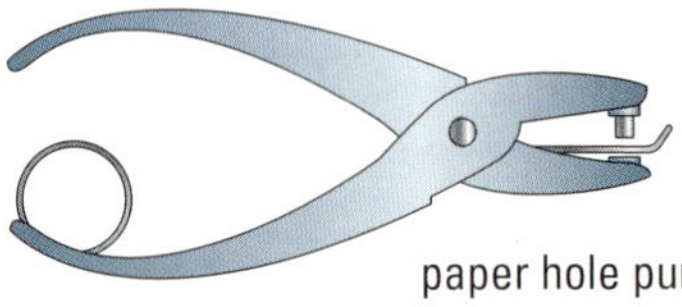

paper hole punch

Structural Parts

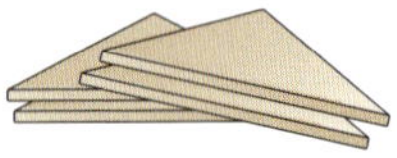

gusset corners

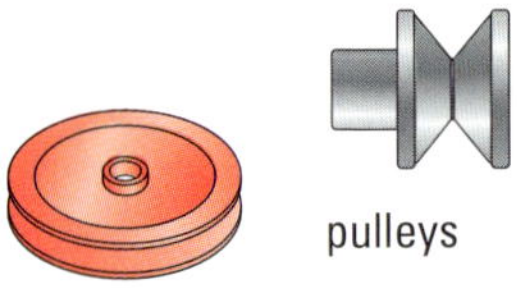

pulleys

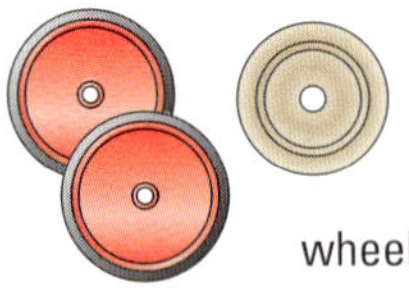

wheels

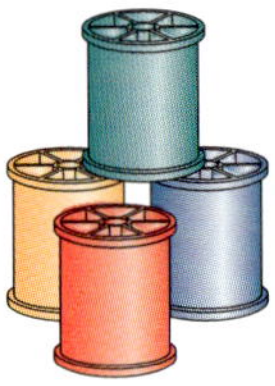

spools

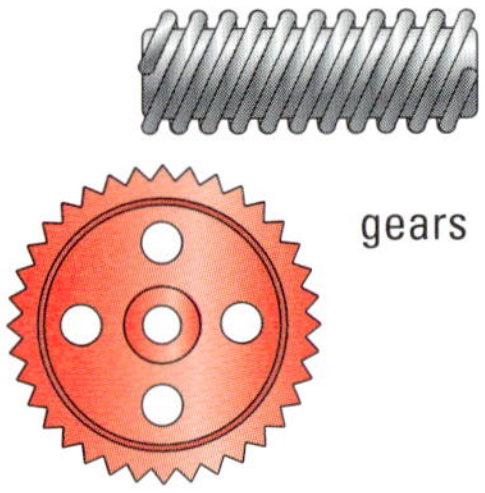

gears

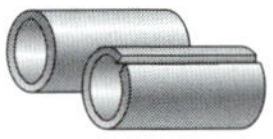

bushings

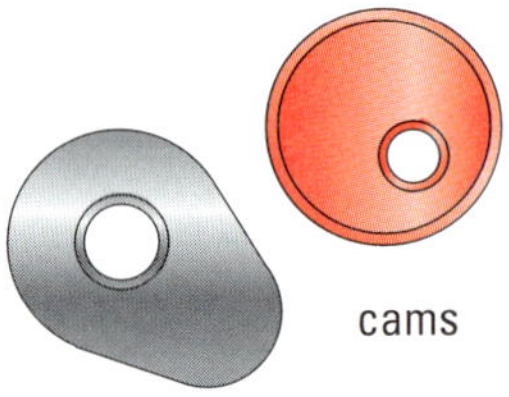

cams

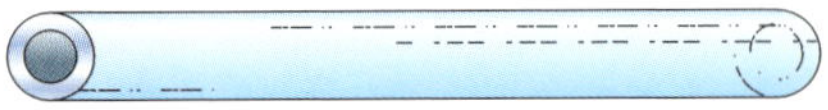

tubing

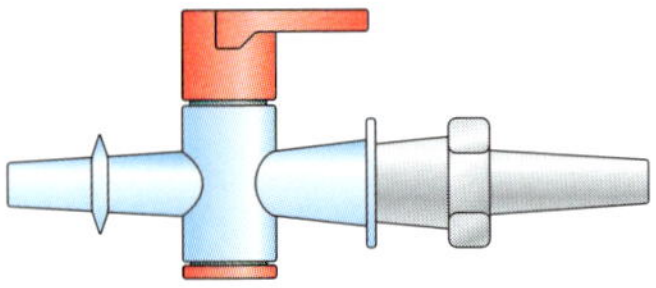

valves

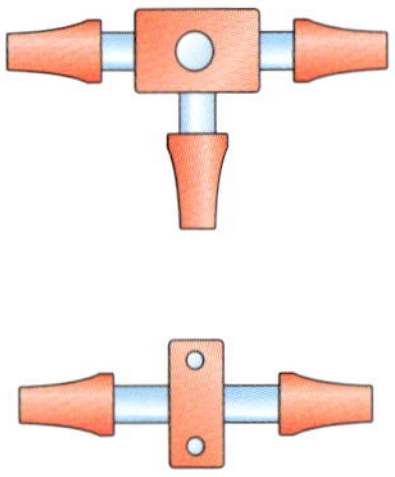

connectors

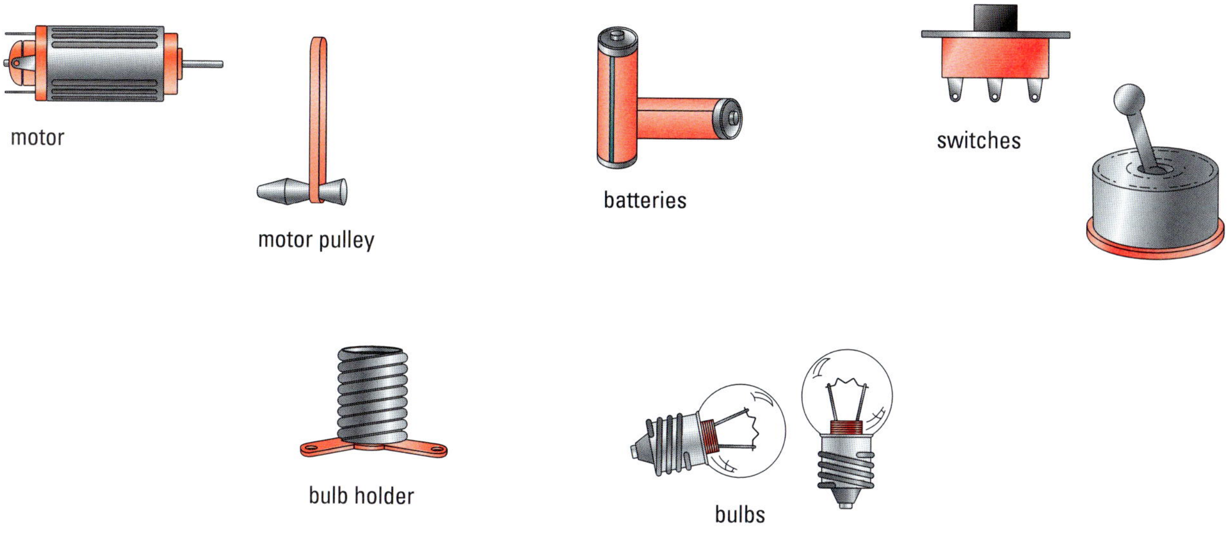

Construction Materials

There are few limits to what materials you can use in construction. The more obvious materials include:

- wood (1 cm x 1 cm basswood)
- bristol board
- metal
- corrugated plastic
- fabric

You should also consider using materials that you encounter everyday, including:

- film containers
- bread tags
- polystyrene food trays
- aluminum pie plates
- fabric
- lids
- pizza "tables"

Try This — Useful Materials

There is a wealth of construction materials available to you. Brainstorm a list of construction materials. Be as creative as possible. For each material, suggest a possible use.

⑤E Fabrication Techniques

There are many different fabrication techniques and skills that you can learn to help take your design from a plan to a prototype. The following section will help you learn some basic techniques. As you become more skilled, you will want to explore additional techniques. You can learn more by taking note of how things are built, by asking for expert help, and by using other information resources.

Structures

The following techniques will help you build a variety of different structures.

Joining with Adhesives

- Use the right adhesive for the materials you are using. Read labels carefully to get the best results and to work safely.
- If possible, it may be useful to roughen the surfaces being joined. A rough surface increases the gluing area and helps the glue hold.
- Make sure the surfaces being joined are in contact with each other. You may need to use a clamp or other device to apply the necessary pressure.

Figure 1 shows how to join two pieces of wood in a right angle. Spread glue on one side of a cardboard triangle, and place it on the wood so that a right angle is formed. When the glue has dried, do the same on the other side of the corner joint. This will prevent sideways twisting and keep the corner at a right angle.

Joining with Fasteners

Use the right fastener for the job. Nails, for instance, are used to join wood permanently whereas screws are better for joining things temporarily. **Figures 2–4** illustrate a few useful joining techniques.

Figure 1
Joining wood in a right angle

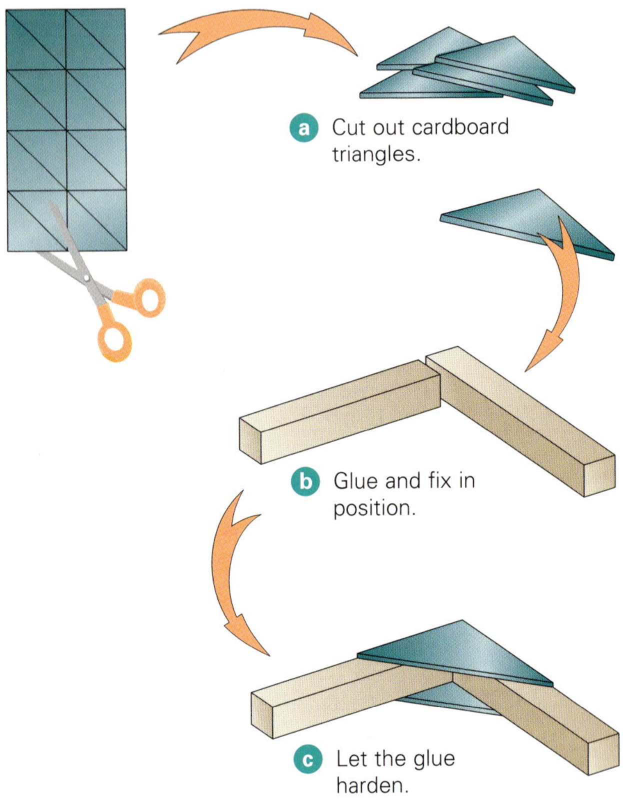

Figure 2
Joining two strips to allow movement

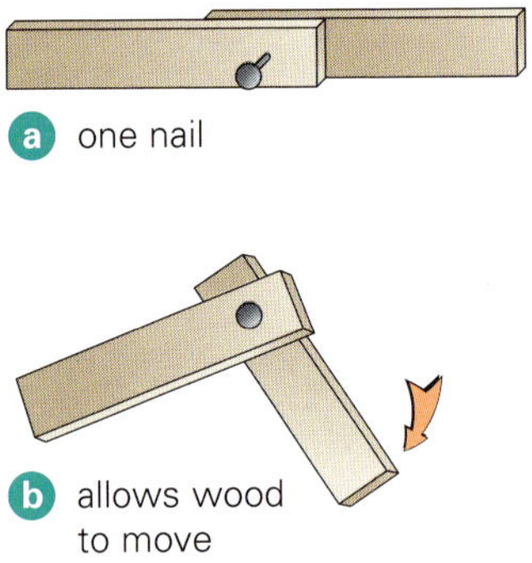

Figure 3
Joining two strips to provide rigidity

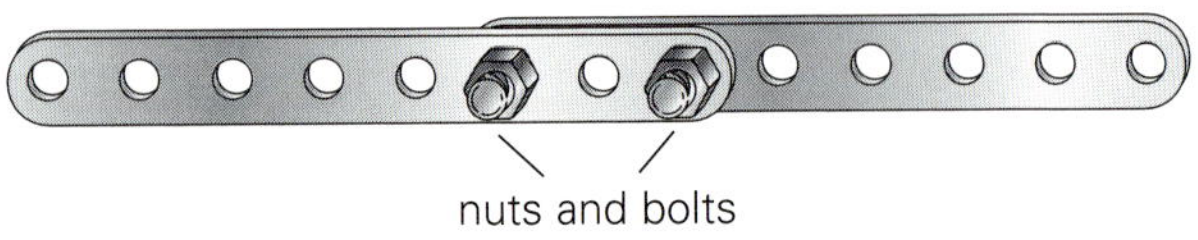

Figure 4

Making holes in materials allows you to use many different fasteners that can help you control friction better. This is useful for constructing such things as levers.

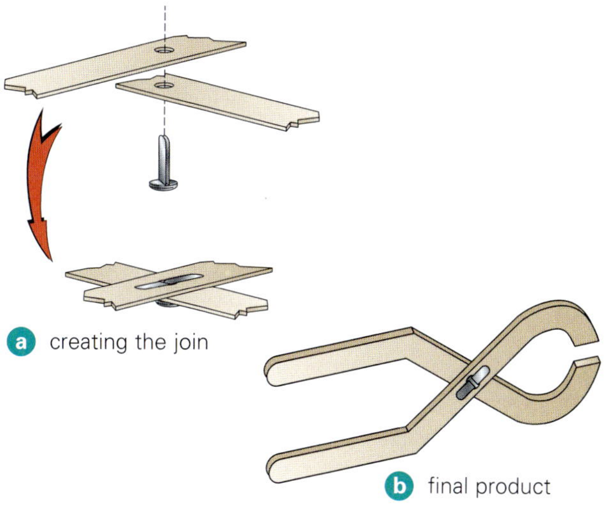

Wheels and Axles

There are two basic kinds of wheel systems: free wheels and fixed wheels. These can be understood by considering a bicycle (**Figure 5**). An axle runs through the centre of every wheel. On a bicycle, the back wheel is fixed to the axle. When the chain turns the axle, the axle turns the wheel. The front wheel is not fixed to the axle—the axle itself is fixed. When the bicycle moves forward, the front wheel turns.

Figure 5

A wheel depends upon friction to make it work. An axle runs through the centre of the wheel.

Free wheel: the bike moves forward, the tire cannot slip and so the wheel turns.

Fixed wheel: the axle turns the wheel, the tire cannot slip and so it pushes the bike forward.

- Wheels must be round. Ways of creating wheels include using ready-made wheels, using round objects as wheels, and cutting dowel into sections.
- Axles must be centred and straight. If not, the ride will be wobbly. Axles must also be loose enough or the wheels will stick.
- Wheels must have enough friction. This can be increased by roughening the surface of the wheel or covering it with a material like rubber.
- Wheels must be well attached to axles. Ways of keeping wheels on axles include plastic or rubber tubing, locking pins, and screws and washers.

Gears

A gear is a wheel with teeth. It can mesh with another gear to create different kinds of motion. Three useful types of gears are illustrated in **Figures 6–8**.

Figure 6

A worm gear drives a cog and turns the axis of rotation through 90°.

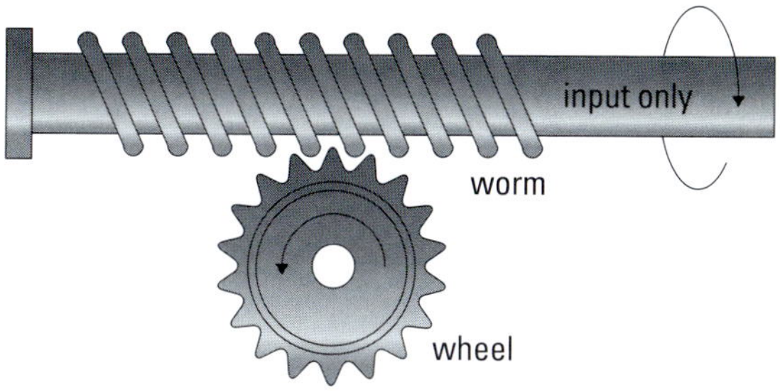

Figure 7

A rack and pinion gear converts circular movement to linear movement.

Figure 8

A bevel gear has teeth cut at a 45° angle. It changes the direction of turning through 90°.

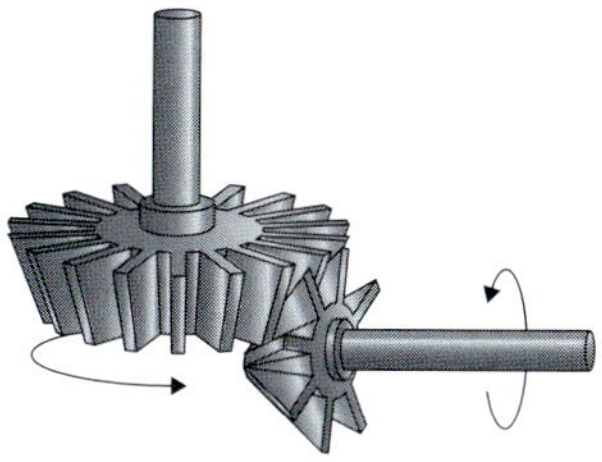

Gears are available ready-made. You can also make simple gears from corrugated cardboard and discs and dowel (**Figure 9**). For gears to work, their teeth must have a common factor. For example, a 12-tooth gear will mesh with a 4-tooth gear because they have a common factor of 3. A 12-tooth gear will not mesh with a 5-tooth gear, for example.

Figure 9
Gear construction

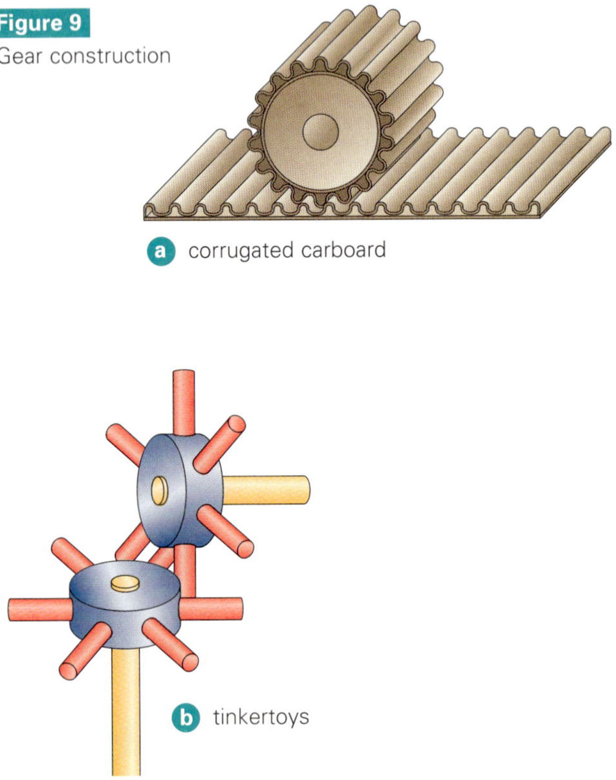

Levers

Levers can be used to change the direction and size of a movement or force. A lever is a rigid rod with a turning point (pivot). There are standard ways of linking levers together that produce specific mechanisms. Some of these are illustrated in **Figure 10**. The best way to understand how levers work is to experiment with simple materials, like cardboard strips and paper fasteners.

Figure 10
Different linkages

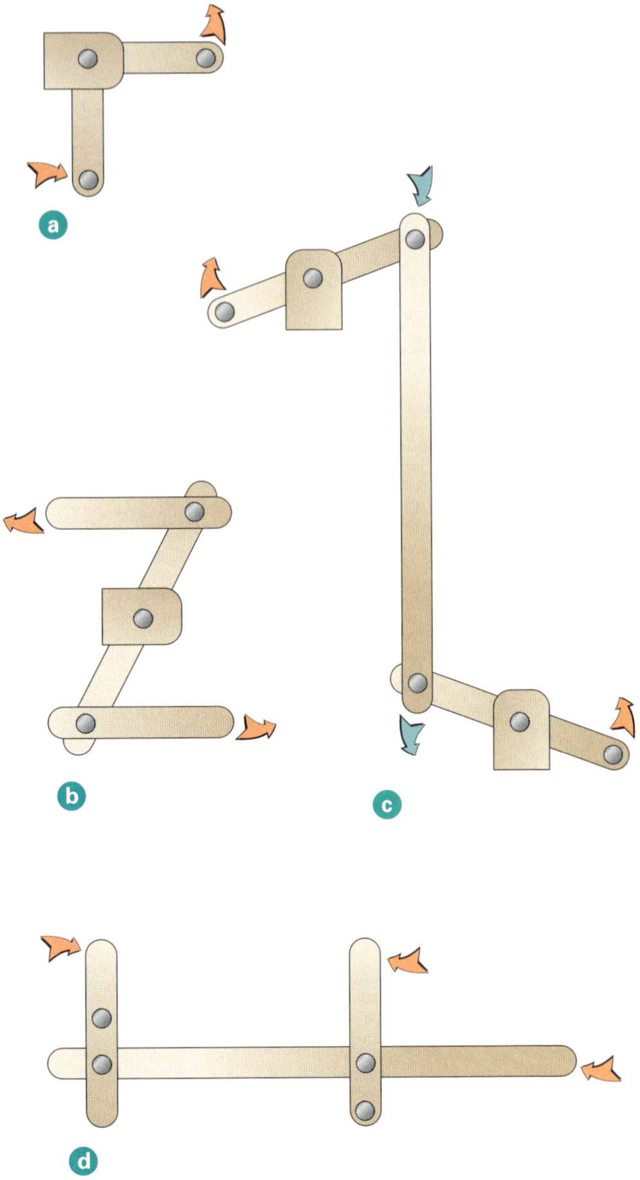

If you are having trouble with your lever, check for the following problems:

- the pivots may be too tight or too loose
- the pivots may not be in the right place
- the linkages may be connected incorrectly
- the linkages may be the wrong length
- the linkages may be out of line.

Electricity

Many of the prototypes you build will use electricity. In these cases, you will need to draw and construct circuits to control electricity. The following section outlines the basic techniques for building circuits. When in doubt about any details, ask your teacher. Think safety first.

Safety Considerations

- Only operate a circuit after it has been approved by your teacher.
- Always ensure that your hands are dry, and that you are standing on a dry surface.
- Do not use faulty dry cells or batteries, do not connect different makes of dry cells in the same battery, and avoid connecting partially discharged dry cells to fully-charged cells.
- Take care not to short-circuit dry cells or batteries.
- Do not use frayed or damaged connectors.
- Handle breakable components with care.

The Parts of an Electric Circuit

Simple electric circuits have four basic parts, shown in **Figure 11**.

Figure 11
The parts of an electric circuit

1. **A source of electrical energy:** You will typically use a combination of dry cells or a special device called a "power supply." These are direct current sources of electricity. They are much safer to use than wall outlets, which provide an alternating current source.
2. **An electrical load:** The load is the appliance that uses the electricity (e.g., lights, buzzers, motors).
3. **A circuit control device:** You will normally use one of a variety of switches (e.g., some switches turn things on and off, some switches reverse the direction of electric motors, etc.)
4. **Connectors:** These are often copper wires.

Drawing a Circuit Diagram

Before building a circuit, you should work it out on paper (**Figure 12**). There are some conventions to follow when drawing circuit diagrams: connecting wires are generally shown as straight linees or 90° angles, and

Figure 12
Schematic diagram of the closed circuit shown in Figure 11

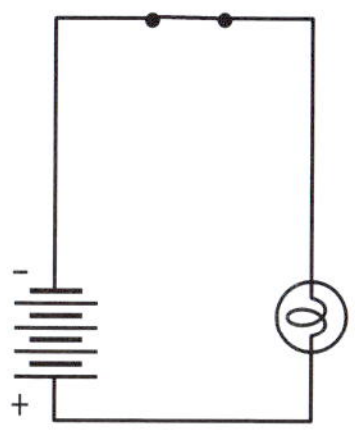

symbols (shown in **Table 1**) are used to represent all the components.

Table 1 **Symbols used in Circuit Diagrams**

	DC CIRCUITS	
Sources/Outlets	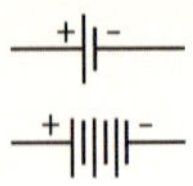	cell
		3-cell battery
Control Devices	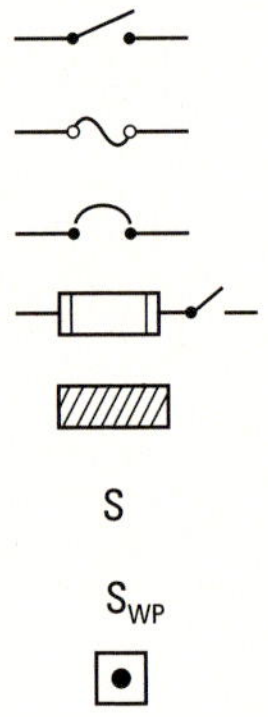	switch
		fuse
		circuit breaker
		switch and fuse
		distribution panel
		switch
		weatherproof switch
		push button
Electrical Loads	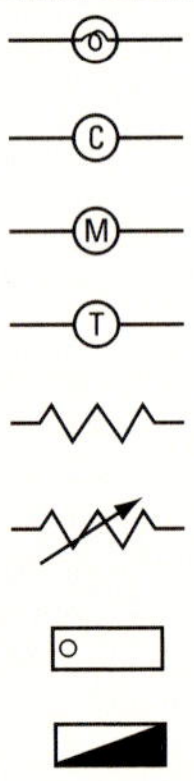	light bulb
		clock
		motor
		thermostat
		resistor
		variable resistor (rheostat)
		fluorescent fixture
		heating panel
Meters	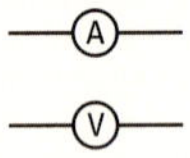	ammeter
		voltmeter
Connectors	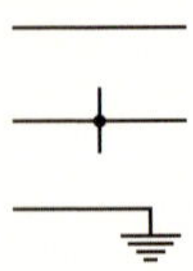	conducting wire
		wires joined
		ground connection

Constructing the Circuit

- Have your circuit diagram approved by your teacher.
- Check connections carefully when linking dry cells. Incorrect connections could cause shorted circuits or explosions. Ask your teacher for clarification if you are unsure.
- When attaching connecting wires to meters, connect a red wire to the positive terminal and a black wire to the negative terminal of the meter. This will remind you to consider the polarity of the meter when connecting it in the circuit.
- Sometimes the ends of connecting wires do not have the correct attachments to connect to the device or meter. Use extra approved attachment devices, such as alligator clips, but be careful to position the connectors so that they cannot touch one another.
- Open the switch before altering a meter connection or adding new wiring or components.
- If the circuit does not operate correctly, open the switch and check the circuit wiring and all connections to the terminals. If you still cannot find the problem, ask your teacher to inspect your circuit again.

Pneumatics and Hydraulics

Two additional systems can be used to create movement: pneumatic systems and hydraulic systems. Each of these systems is based on the fact that fluids (gas or liquid) can be "pushed" in order to make something else move. Hydraulic systems use a liquid. Water, for instance, works in low-pressure systems, whereas hydraulic oil is used in high-pressure systems. Pneumatic systems use a gas, such as air. The air must be compressed before it can be used.

These systems have three basic components:

1. Inputs: You need a reservoir of fluid and a way of moving the fluid into the system. For instance, gravity can be used to create water pressure.
2. Controls: You need a way of directing the flow of the fluid that is strong enough to withstand the pressure of the fluid. The parts of the control system include pipes to control the direction of the flow, connectors to extend the system, taps to control the amount of fluid in the system, and valves to control the direction of the flow.
3. Outputs: The moving fluid is used to activate a device.

A simple pneumatic system is illustrated in **Figure 13**. Air from the syringe is forced through the tubing and into the balloon. The balloon inflates and the hand is forced open.

Figure 13
A pneumatic hand

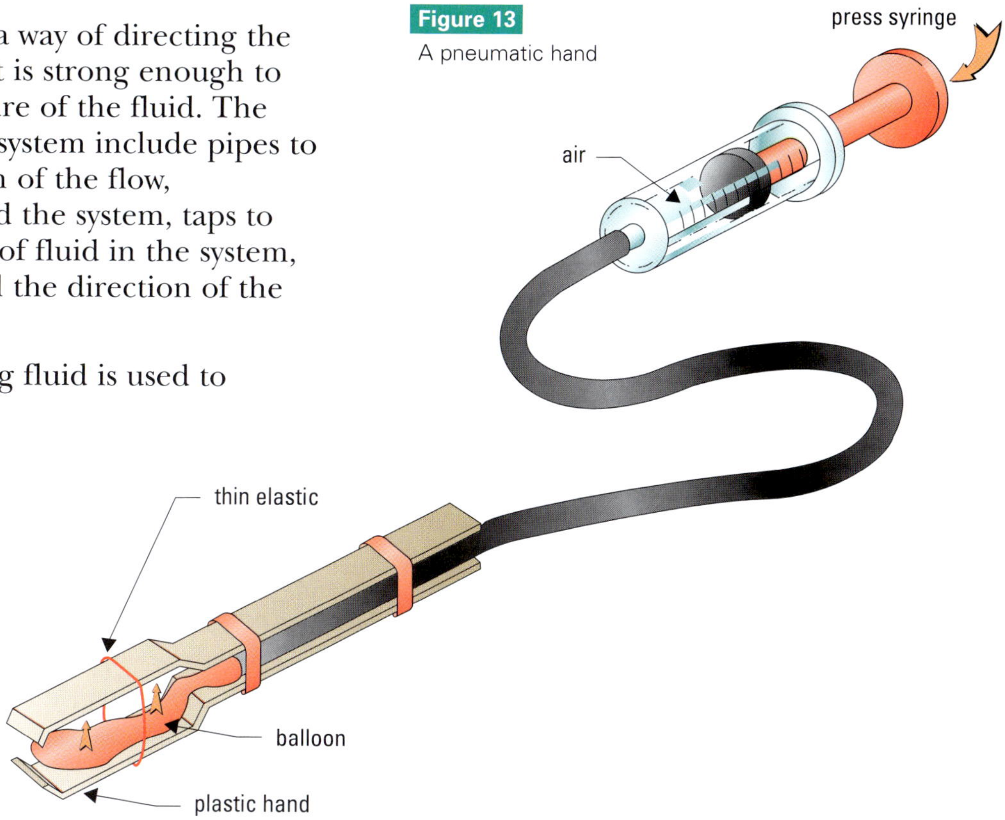

6 Observing and Recording Data

6A Obtaining Qualitative Data

Qualitative Observations

An observation is information that you get through your senses. Scientific questions and technological problems are usually based on observations. For instance, watching waves wash up onto a beach might make you wonder about the cause of tides or how to prevent soil erosion. When you describe the qualities of objects, events, or processes, the observations are qualitative (**Figure 1**).

Figure 1
The shape, the space available for sitting, and the colour of the chair are qualitative observations.

Making observations is a critical step in the process of scientific inquiry (see 2A "Process of Scientific Inquiry," Step 6). Making observations is also critical to the process of design. Evaluating a prototype according to its design criteria requires careful observation (see 3A "The Problem-Solving Cycle," Step 5). Both scientists and technologists have grouped qualitative observations into categories, based on the kind of qualities displayed and the purpose of the investigation.

Qualitative Observations in the Process of Scientific Inquiry

Common categories of qualitative observations used in the process of scientific inquiry appear below:

State of Matter: All substances can be classified as solid, liquid, or gas (**Figure 2**).

Figure 2

a Roller blades are solid.

b Water is a liquid.

c Air is a gas.

Colour: Objects can be described as being any colour or any shade of colour (**Figure 3**). Materials that have no colour should be described as colourless.

Figure 3
This leaf appears green because it reflects green light back to your eyes.

Smell: Also known as odour. There are many words to describe smells, including pungent, strong (**Figure 4**), spicy, sweet, and odourless.

Figure 4
Skunks can produce a very strong odour.

Texture: The surfaces of objects can have a variety of textures, including smooth, rough, prickly, fine, and coarse (**Figure 5**).

Figure 5
These cereal flakes have a coarse texture.

Taste: Objects can taste sweet, sour, bitter, or salty (**Figure 6**). Other tastes are combinations of these basic tastes. Objects that have no taste can be described as tasteless.

Figure 6
Black olives give this pizza a salty taste.

Shininess: Also known as lustre. Objects with very smooth surfaces that reflect light easily, like mirrors, are said to be shiny or lustrous (**Figure 7**). Kitchen taps, mirrors, even well polished desktops can be described as lustrous. Objects with dull surfaces are said to be non-lustrous.

Figure 7
Most metals, such as copper and silver, appear shiny.

Clarity: Some substances let so much light through that letters can be read through them. These substances are said to be clear or transparent. Other substances that allow light through, but not in a way that allows you to see through them, are translucent. Objects that do not let light through are opaque (**Figure 8**).

Figure 8
Clarity ranges from transparent to opaque.

Other qualitative descriptions include form (the shape of a substance), hardness, brittleness (how easily the substance breaks), malleability (the ability of the object to be changed into another shape), and viscosity (a liquid's resistance to flow).

Another important characteristic that can be described qualitatively is the ability of substances to combine with each other.

Qualitative Observations in the Process of Design

The following list outlines design elements that often form the basis for qualitative observations in the process of design:

Shape: A shape is an area defined by a line. A shape can have two dimensions, like a square, or three dimensions, like a pyramid (**Figure 9**). The three dimensions are called: height, depth, and width.

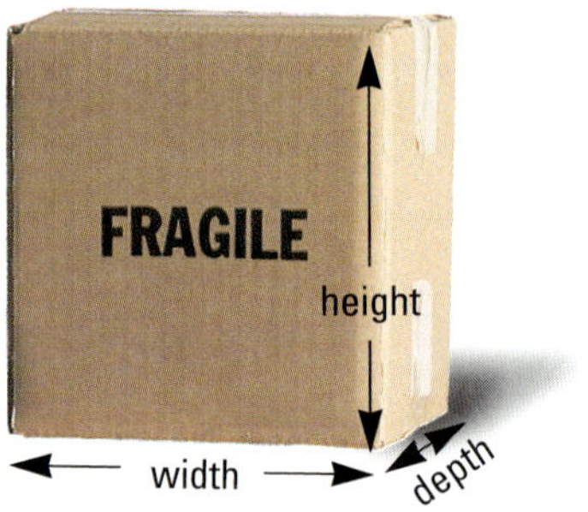

Figure 9
The surface of one side of this box is a square. The box itself is a cube.

Space: Space is the area that surrounds an object or is contained within an object. All objects exist in space (**Figure 10**). We can only appreciate the form of an object by imagining or seeing the space around it.

Figure 10
The shapes of the objects in this room can only be understood because of the space around them.

Colour: Colour is important because it produces an instant response. Colour is used to enhance a design.

Texture: Texture has an effect on the appearance and function of a design.

Ergonomics: Ergonomics involves designing things so that they can be used easily and safely.

Try This — Which Qualities?

Review the categories of qualities used in the processes of scientific inquiry and design. You will notice that some appear in both lists. Are there others that you think could be useful in both processes? Provide two or three examples.

6B Obtaining Quantitative Data

Quantitative Observations

Observations that are based on measurements or counting are said to be quantitative, since they deal with quantities of things. The length of a rose's stem or a piece of wood, the number of petals on a flower, and the number of rotations of a gear are all quantitative observations.

Look at the two lines in **Figure 1**. Which looks longer? You will find that AB and CD are the same length. Our senses can be fooled. That is one of the reasons quantitative observations are important in science and technology and it is also the reason measurements must be made carefully.

Figure 1
Which line is longer? Use a ruler.

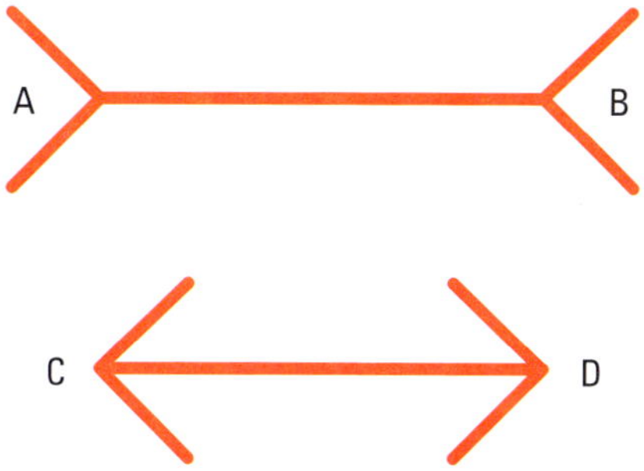

Standards of Measurement

Units of measurement used to be based on local standards that the community had agreed to. At one time, for example, there was a unit equal to the length of a line of 16 people standing close together. Horse heights are still measured in hands, based on the width of a human hand, and measured from the ground to the horse's shoulder.

These standards may sound strange, but the unit of length that replaced most local standards, the metre, was based on an arc on the Earth that ran from the equator, through Barcelona, Paris, and Dunkirk, to the North Pole (**Figure 2**). The length of this arc, divided by 10 000 000, equaled one metre. That also sounds strange, but it established the first standard unit of length that the whole world could use.

Figure 2
The early standard for the metre was 1/10 000 000 of the distance from the equator to the North Pole.

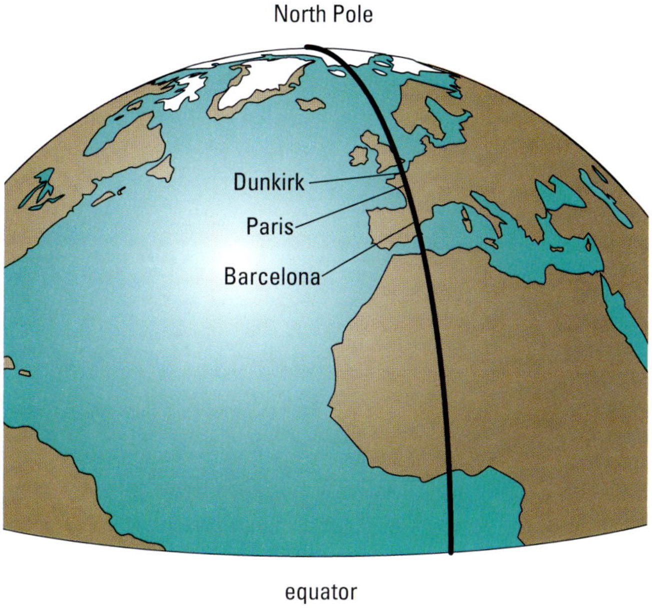

The metric system, which includes units such as the metre and the kilogram, is the one adopted by Canada. You should be familiar with, and use, the units from this international system (also called the SI system, from the French name: Système International d'Unités).

Base Units and Prefixes

There are seven SI base units, shown in **Table 1**.

Table 1 **The Seven SI Base Units**

Quantity	Unit	Symbol
length	metre	m
mass	kilogram	kg*
time	second	s
electric current	ampere	A
temperature	kelvin	K
amount of substance	mole	mol
light intensity	candela	cd

*The kilogram is the only base unit that contains a prefix. The gram proved to be too small for practical purposes.

Larger and smaller units are created by multiplying or dividing the value of the base units by multiples of 10. For example, the prefix deca means multiplied by 10. Therefore, one decametre (1 dam) is equal to ten metres (10 m). The prefix kilo means multiplied by 1000, so one kilometre (1 km) is equal to one thousand metres (1000 m). Similarly, each unit can be divided into smaller units. The prefix milli, for example, means divided by 1000, so one millimetre (1 mm) is equal to 1/1000 of a metre.

To convert from one unit to another, you simply multiply by a conversion factor. For example, to convert 12.4 metres to centimetres, you use the relationship 1 m = 100 cm.

$$12.4 \text{ m} = ? \text{ cm}$$

$$12.4 \text{ m} \times \frac{100 \text{ cm}}{1 \text{ m}} = 1240 \text{ cm}$$

To convert 6.3 g to kilograms, you use the relationship 1000 g = 1 kg.

$$6.3 \text{ g} = ? \text{ kg}$$

$$6.3 \text{ g} \times \frac{1 \text{ kg}}{1000 \text{ g}} = 0.0063 \text{ kg}$$

Any conversion of the same physical quantities can be done in this way. The conversion factor is chosen so that, using cancellation, it yields the desired unit.

Once you understand this method of conversion, you will find that you can simply move the decimal point. Move it to the right when the new unit is smaller and to the left when the new unit is larger. As you can see from **Table 2**, not all the units and prefixes are commonly used.

Table 2 **Metric Prefixes**

Prefix	Symbol	Factor by which the base unit is multiplied	Example
giga	G	10^9 = 1 000 000 000	
mega	M	10^6 = 1 000 000	10^6 m = 1 Mm
kilo	k	10^3 = 1 000	10^3 m = 1 km
hecto	h	10^2 = 100	
deca	d	10^1 = 10	
		10^0 = 1	m
deci	da	10^{-1} = 0.1	
centi	c	10^{-2} = 0.01	10^{-2} m = 1 cm
milli	m	10^{-3} = 0.001	10^{-3} m = 1 mm
micro	μ	10^{-6} = 0.000 001	10^{-6} m = 1 μm

Table 3 shows quantities that you may need to be familiar with.

Table 3 **Common Quantities and Units**

Quantity	Unit	Symbol
length	kilometre	km
	metre	m
	centimetre	cm
	millimetre	mm
mass	tonne (1000 kg)	t
	kilogram	kg
	gram	g
area	hectare (10 000 m2)	ha
	square metre	m^2
	square centimetre	cm^2
volume	cubic metre	m^3
	litre	L
	cubic centimetre	cm^3
	millilitre	mL
time	minute	min
	second	s
temperature	degrees Celsius	°C
force	newton	N
energy	kilojoule	kJ
	joule	J
pressure	kilopascal	kPa
	pascal	Pa

Measuring Accurately

Many people believe that all measurements are accurate and dependable. But there are many things that can go wrong when measuring. The instrument may be faulty. Another similar instrument may give different readings. The person making the measurement may make a mistake.

Ask your teacher for two thermometers and check the room temperature. Are the readings identical? (You may want to check both thermometers over a range of temperatures—do the two thermometers always give the same readings?) Suppose you use only one thermometer, and it reads 21°C. Is the room temperature really 21°C? That depends on how reliable the thermometer is. If your thermometer is not accurate, the temperature of the room could be 19°C, 23°C, or some other value.

No matter what you are measuring, it is wise to repeat your measurement at least three times in order to be sure of an accurate result. If your measurements are close, calculate the average and use that number. To be more certain, repeat the measurements with a different instrument.

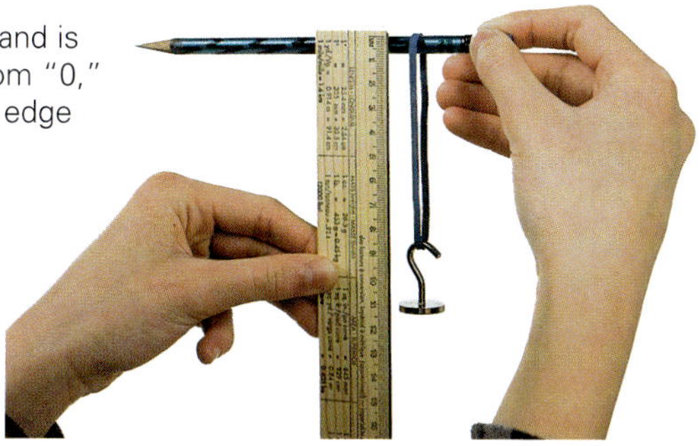

Figure 4
The rubber band is measured from "0," not from the edge of the ruler.

Measuring Tips

Measuring temperature: If you are measuring the temperature of a liquid, keep the bulb of the thermometer near the middle of the liquid (**Figure 3**). If the liquid is being heated and the thermometer is simply sitting in the container with its bulb at the bottom, you will be measuring the temperature of the bottom of the container, not the temperature of the liquid. (Do not use the thermometer as a stirring stick!) Similarly, if you are measuring the temperature of a solid, place the bulb of the thermometer as close to the centre of the solid as possible.

Figure 3
The thermometer is not resting on the bottom of the beaker.

Measuring length: Choose the right measuring instrument for the task at hand. If you are measuring something quite short, for instance, you need a tool that is quite precise (i.e., that indicates small increments.) Make sure you start measuring where "0" is indicated (**Figure 4**). Many rulers or metre sticks extend past this point.

Measuring construction materials: Remember that when you cut some materials (e.g., wood) you will lose a bit of wood in cutting (**Figure 5**). This "bit" must remain the same to ensure square corners or exact angles. It will also affect how large the final pieces of your product are.

Figure 5
You will sometimes lose material when you cut it.

Measuring volume: As with length, it's important to choose the right measuring instrument. Make sure your measuring instrument is on a level surface. If you are measuring a solid, shake the container so that the surface is level. To measure the volume of a liquid using a graduated cylinder, measure from the bottom of the apparent curve (**Figure 6**).

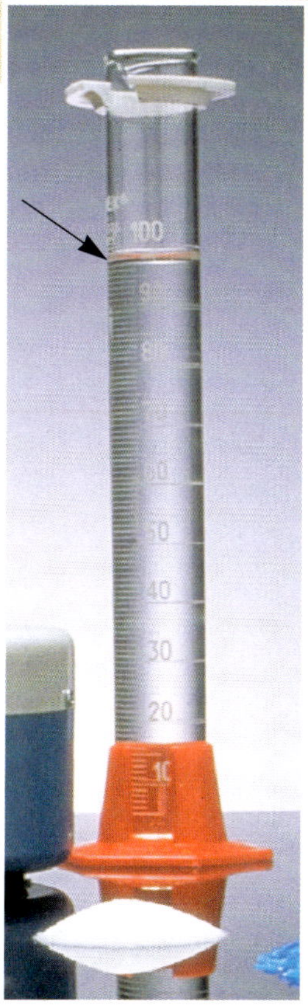

Figure 6
The arrow indicates where the measurement should be taken.

Measuring time: Measuring short intervals of time introduces a strong possibility of human error (**Figure 7**). If you are using a stopwatch to time how many times a pendulum swings in 5 s, for instance, you have to start the stopwatch at precisely the same moment as the pendulum begins its first swing. Repeating the process and averaging the results is advisable.

Figure 7
Accurate timing requires a lot of attention.

Measuring mass and weight: Refer to 5B "Working with Scales and Balances."

Try This: Measuring Quantities

Complete one of the following activities.

Activity A:

- You have been given a bag of balloons to provide motion for the balloon-powered car you have designed. Each balloon is advertised as being the same diameter. You want to verify this so that you can be certain all groups use the same volume of air to power their vehicles. Check to make sure that this is actually true.
- Decide how your group will prove or disprove this.
- Record all quantitative data in a table. What is the average balloon size?
- Compare your results with other groups.

Activity B:

- Find out what pH level means and the scale used to identify pH levels as acidic, neutral, and alkaline.
- With a partner, measure the pH levels of a variety of liquids.
- Compare your results with others in your class. Did everyone obtain the same results?

6C Scientific & Technical Drawing

Scientific drawings are done to record observations as accurately as possible. Technical drawings are used to visualize design ideas. Drawings are used to communicate, which means they must be clear, well-labelled, and easy to understand. Following are some tips that will help you produce useful scientific and technical drawings.

Before You Begin

- **Obtain some paper.** You will have to make a decision about which type of paper best meets your needs. Blank paper will be best when lines might obscure your drawing or make your labels confusing. Graph paper will be best when you are drawing something that requires exact measurements.
- **Find a sharp, hard pencil** (e.g., 2H or 4H). Avoid using pen, thick markers, or coloured pencils. Ink can't be erased—even the most accomplished artists change their drawings—and coloured pencils are soft, making lines too thick.
- **Plan your drawing.** Ensure that your drawing will be large enough that people can see details. For example, a third of a page might be appropriate for a diagram of a single cell or a gear train. If you are drawing the entire field of view of a microscope, draw a circle with a reasonable diameter (e.g., 10 cm) to represent the field of view. If you are drawing a three-dimensional object, plan for more than one view (e.g., top view, side view, etc.)
- **Leave space for labels.** In scientific drawings, the right side of the drawing is commonly used.

Scientific Drawing Tips

- **Make simple, two-dimensional drawings.**
- **Draw only what you see.** Your textbook may act as a guide, but it may show structures that you cannot see in your specimen.
- **Do not sketch.** Draw firm, clear lines, including only relevant details that you can see clearly.
- **Do not use shading or colouring.** A stipple (series of dots) shown in **Figure 1** may be used to indicate a darker area. Use double lines to indicate thick structures.

Figure 1
Using a stipple effect

Labelling

- **Label all drawings fully.** Avoid printing labels directly on the drawing.
- **Use a ruler.** Label lines must be horizontal and ruled firmly from the area being identified to the label (**Figure 2**).

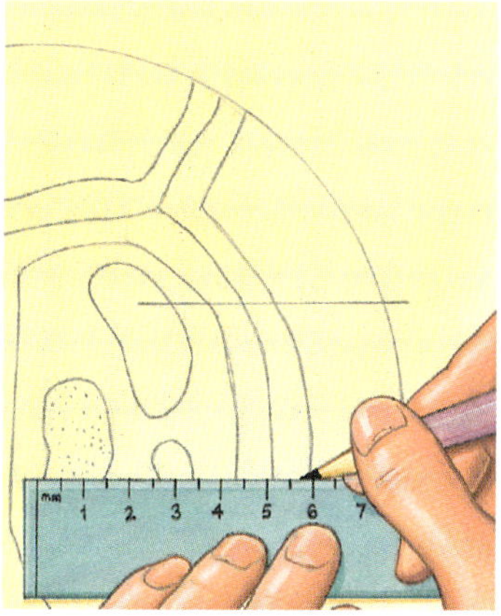

Figure 2
Neat labelling makes your drawing clear.

- **Label lines should never cross.**
- **Label drawings neatly.** In scientific drawings, it is preferable to list your labels in an even column down the right side.
- **Title the drawing.** Use the name of the specimen or prototype and any other specific information that will help identify the drawing. Underline the title.

Scale Ratio

To show the relation of the actual size to your drawing size, print the scale ratio of your drawing beside the title.

$$\text{scale ratio} = \frac{\text{size of drawing}}{\text{actual size of the specimen/prototype}}$$

For example, if you have drawn a nail that is 2 cm long (**Figure 3**) and the drawing is 8 cm long, then the scale ratio, which in this case is a magnification, is

$$\frac{8 \text{ cm}}{2 \text{ cm}} = 4\times$$

Figure 3

The nail as drawn is 4X larger than the actual nail.

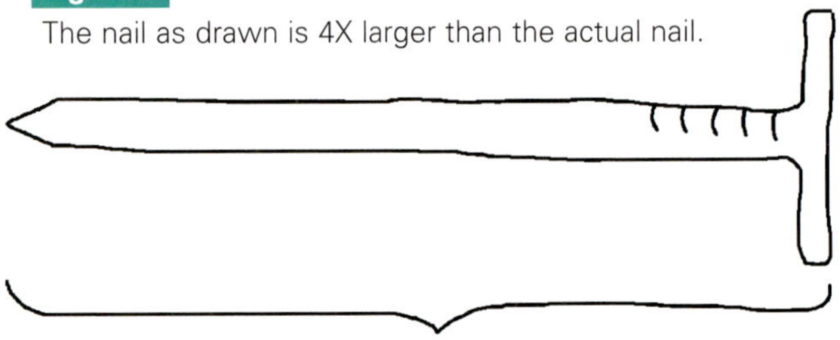

The magnification is always written with an "×" after it. In a fully labeled drawing, the total magnification of the drawing should be placed at the bottom right side of the diagram. If the ocular lens magnified a specimen 10×, the low-powered objective (4×) was used, and the diagram was drawn 3× larger that the original specimen, the total magnification of the diagram would be as follows:

$$\begin{aligned}\text{Total Magnification} &= \text{Ocular Lens} \times \text{Low-Powered Objective Lens} \times \text{Scale Ratio}\\ &= 10 \times 4 \times 3\\ &= 120\times\end{aligned}$$

The total magnification should be written on the bottom right-hand side of the diagram, as shown in **Figure 4**.

Figure 4

The magnification is 120×.

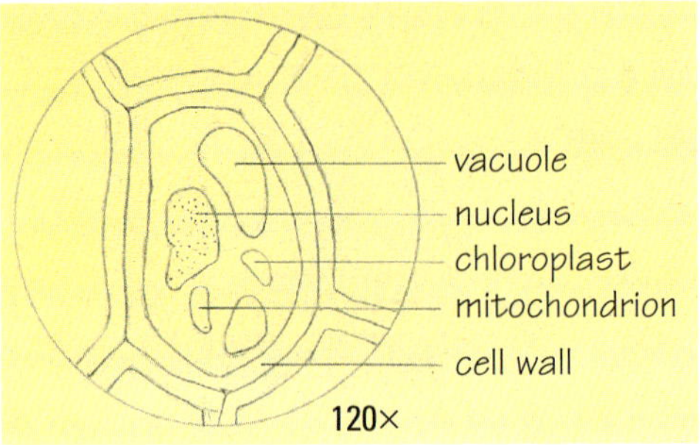

Technical Drawing Techniques

The kind of drawing you produce during the process of technological design will depend on its purpose. Some drawings are little more than doodles—rough attempts to give shape to your ideas at the beginning of the process. Some drawings will help you explain your concept to others. Some drawings are used in the plan of a prototype, along with written instructions, to help in the construction.

Sketches

When you are first brainstorming ways of solving a problem, it is often useful to sketch your ideas. The sketches help you develop and record your concept. Often a sketch will include notes to help you keep track of an evolving idea (**Figure 5**). Sketches can also help you communicate your ideas to other members of your team.

Figure 5

A sketch for a game design

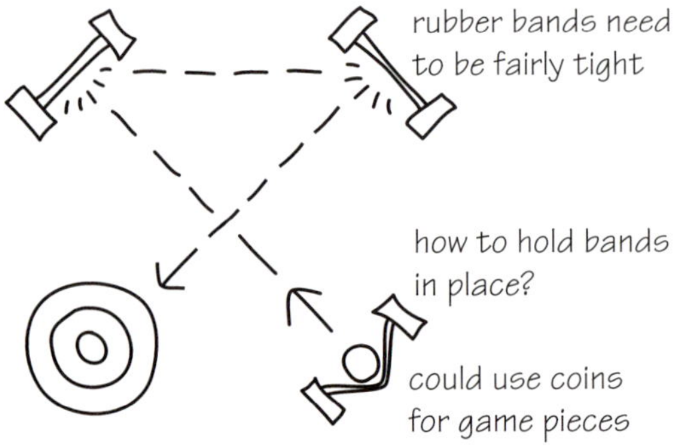

Isometric Drawings

As you refine your design, you will also refine your technical drawings. Drawings that are designed to help others understand or build your design need to be clear and accurate.

An isometric drawing represents an object in three dimensions with angles drawn at 30°. Often special isometric grid paper is used. It consists of three sets of parallel lines. Wherever the lines cross, there are three axes which represent depth, length, and height (**Figure 6**). This method of drawing allows you to use the actual measurements of the object you are drawing and provides others with a picture of the object that it easy to understand (**Figure 7**).

Figure 6
The parallel lines in an isometric grid run at 60° to one another.

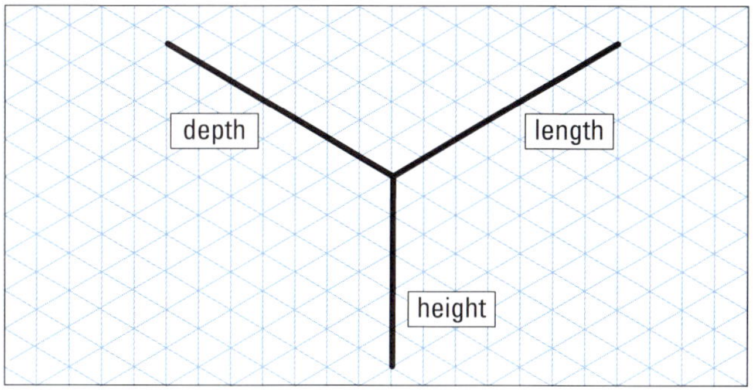

Figure 7
A model of a garage to be made from a shoebox

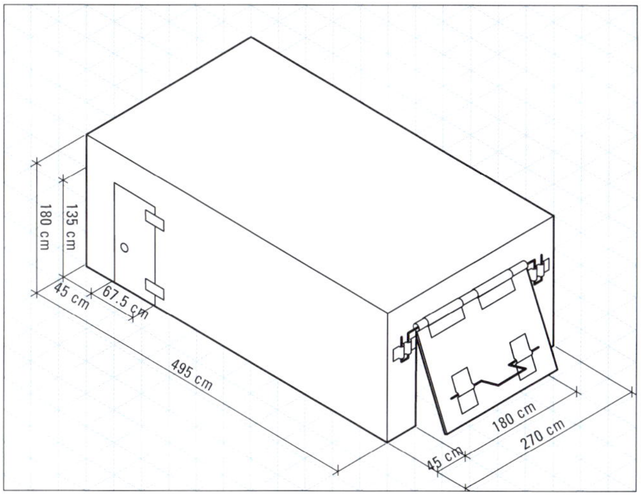

Orthographic Projection

It is often useful to create a working drawing that shows your prototype from three different views: the top, side, and front views. An orthographic projection is an accurate drawing that shows all the views and includes every detail and measurement (**Figure 8**). Each view is drawn square-on, in two dimensions. It is drawn so that the relation between the views is clear. Using graph paper makes this easier. One way of understanding the relation between each view is to imagine that the object is enclosed in a transparent box. If you drew what you saw from sides A, B and C on the box, then opened the box up, you could see how each view is related to the other two (**Figure 9**).

A complete working drawing often includes the orthographic projection, a list of parts, a list of materials used for each part, and detailed assembly instructions.

Figure 8

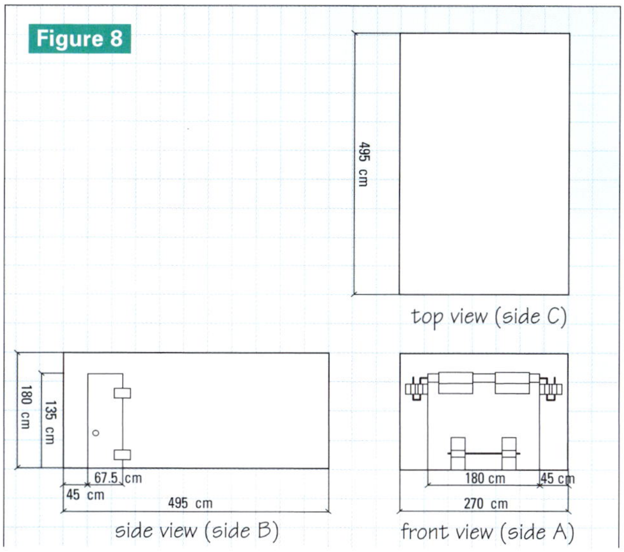

Figure 9

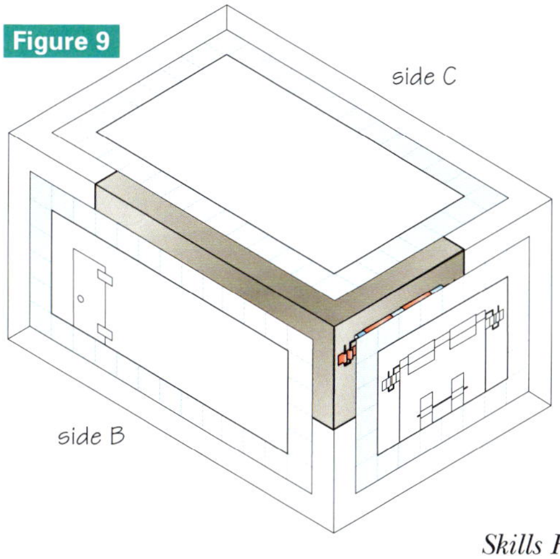

Other Views of an Object

In some cases, it may be useful to show the inside of an object. One way of doing this is to draw a cross-section (**Figure 10**).

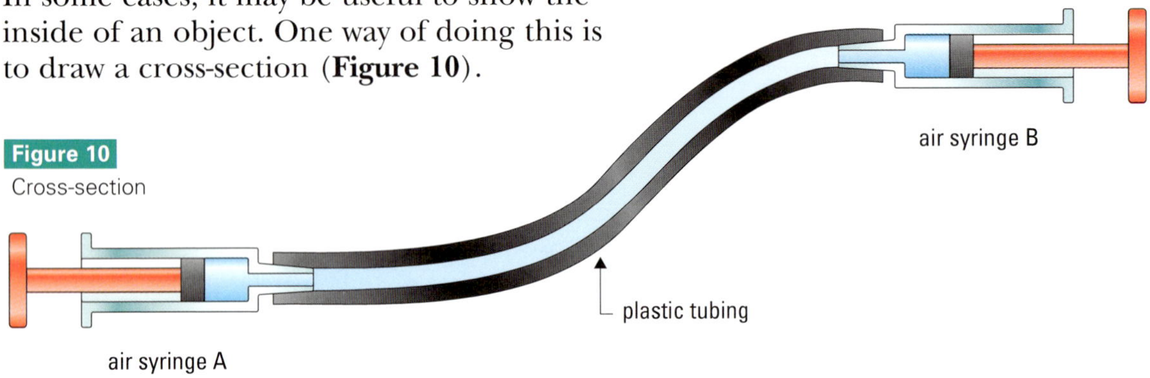

Figure 10
Cross-section

To show clearly how parts of an object fit together, an exploded view is often drawn (**Figure 11**).

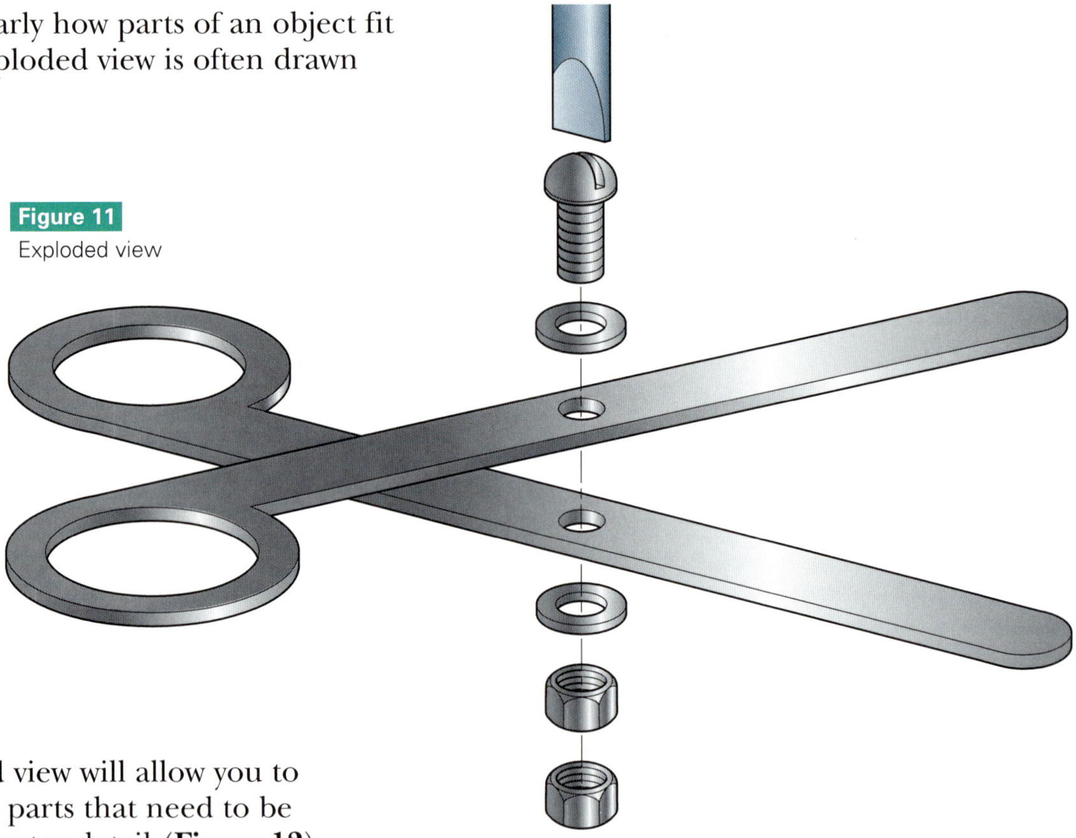

Figure 11
Exploded view

An enlarged view will allow you to represent small parts that need to be explained in greater detail (**Figure 12**).

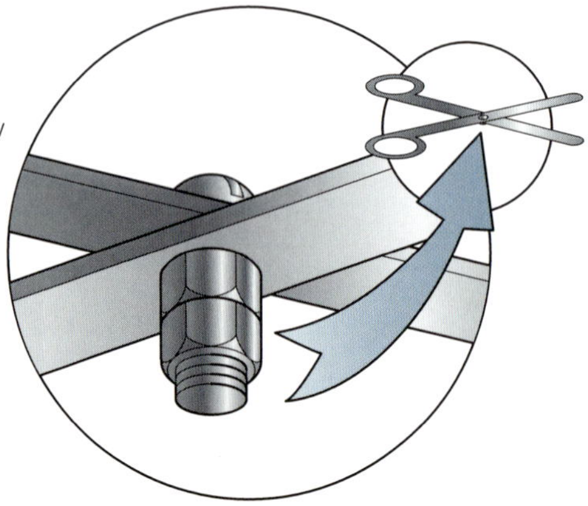

Figure 12
Enlarged view

6D Creating Data Tables

Creating effective data tables in your investigations will help you record and analyze your data. Constructing a useful data table is one of the first steps in making sense of your experimental data. Take a look at **Tables 1, 2,** and **3** which come directly from *Nelson Science & Technology 7/8*. What similarities exist? What strategies should you employ when constructing your data tables?

Table 1 Maturity Time and Yield of Tomato Varieties

Variety	Time to maturity (days)	Yield for determinate varieties (kg per plant)	Yield for indeterminate varieties (kg per plant every 10 d)
A	80	n/a	6.8
B	40	n/a	2.6
C	70	n/a	5.2
D	60	10	n/a
E	70	n/a	4.2
F	65	10	n/a
G	55	15	n/a
H	50	10	n/a
I	75	n/a	5.6

Table 2 Average Monthly Temperatures (°C) in Cities A and B

Month	Temperature (°C) in City A	Temperature (°C) City B
J	-7	-6
F	-6	-6
M	-1	-2
A	6	4
M	12	9
J	17	15
J	21	18
A	20	18
S	15	14
O	9	9
N	3	3
D	-3	-3

Table 3 Average Monthly Precipitation (cm) in Cities A and B

Month	Precipitation (cm) in City A	Precipitation (cm) in City B
J	4.6	14.7
F	4.6	11.9
M	5.7	12.3
A	6.4	12.4
M	6.6	11.1
J	6.9	9.8
J	7.7	9.7
A	8.4	11.0
S	7.4	9.5
O	6.3	12.9
N	7.0	15.4
D	6.6	16.7

The following checklist will help you construct effective data tables in your investigations:

- List the dependent variable(s) (the effect) along the top of the table.
- List the independent variable (the cause) along the side of the table.
- Be sure that each data table has a descriptive, yet concise, title.
- Be sure to include the units of measurement along with each variable when appropriate.
- If you include the results of your calculations in a table, be sure to show at least one sample calculation in your data analysis.

Refer to 9C "Using Your Computer Effectively" for information about using spreadsheet programs to create useful data tables.

7A The Need to Graph

Making Sense of Data

Scientists and technologists often create huge amounts of data while doing experiments and studies—maybe hundreds, even thousands, of numbers for every variable. How can this mass of data be arranged so that it is easy to read and understand? That's right—in a graph. The sample tables below don't have thousands of pieces of data, but there is enough to become confusing. Can you make sense of the data in **Table 1** by simple inspection?

A graph is an easy way to see where a relationship or pattern exists. As well, it allows you to see more precisely what the relationship is, so it can be accurately described in words and by mathematics. **Figure 1** is a point-and-line graph that shows the data from **Table 1**.

The graph in **Figure 1** shows the relationship between the two variables and highlights two temperature plateaus where a change of state is taking place. In more complex relationships such as this, the need for a graph is quite clear. Data trends are much easier to visualize and understand in the organized form of a graph than as numbers in a table.

Table 1 The Heating Curve of Water

Time (minutes)	Temperature (°C)	Time (minutes)	Temperature (°C)
1	0	13	51
2	0	14	58
3	0	15	65
4	2	16	72
5	4	17	79
6	7	18	86
7	9	19	93
8	16	20	97
9	23	21	100
10	30	22	100
11	37	23	100
12	44	24	100

Figure 1 The Heating Curve of Water

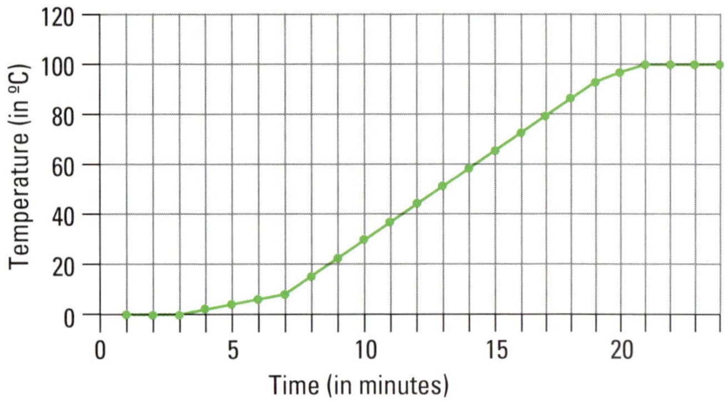

Types of Graphs

There are many types of graphs that you can use when organizing your data. You must identify which type of graph is best for your data before you start drawing it. Three of the most useful kinds are point-and-line graphs, bar graphs, and circle graphs (also called pie graphs).

Point-and-Line Graphs

When both variables are quantitative, use a point-and-line graph. The graph in **Figure 2** was created after an experiment that measured the number of worms on the surface of soil (quantitative) and the volume of rain that fell on the soil (quantitative). It is based on the data in **Table 2**.

Table 2 Number of Worms Per Volume of Water

Volume of water (mL)	Number of Worms
0	3
10	4
20	5
30	9
40	22

Figure 2

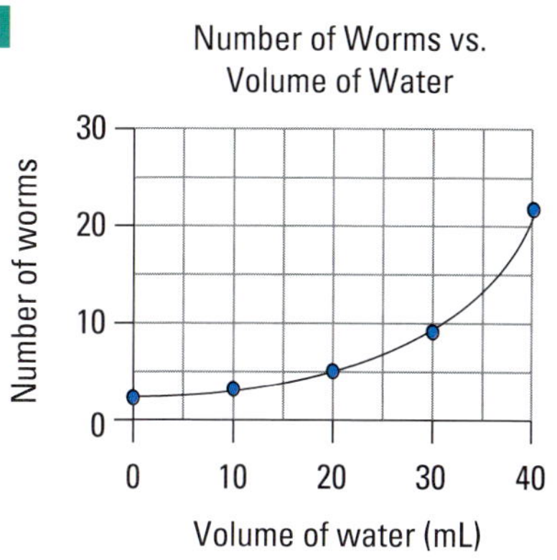

Bar Graphs

When at least one of the variables is qualitative, use a bar graph. For example, a study of the math marks of students (quantitative) who listened to different kinds of music (qualitative) while doing their math homework resulted in the graph in **Figure 3**. In this kind of graph, each bar stands for a different category, in this case a type of music. Notice also that the range on the vertical axis is chosen so that even the smallest bar is still visible.

Figure 3

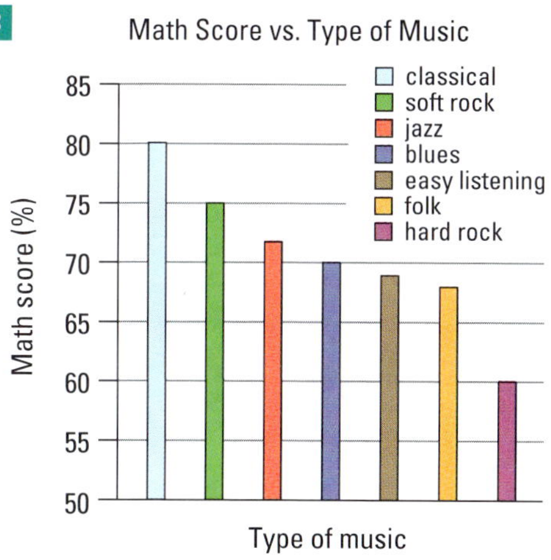

Circle (or Pie) Graphs

Circle graphs and bar graphs are used for similar types of data. If your quantitative variable can be changed to a percentage of a total quantity, then a circle graph is useful. For example, if you surveyed a class to find the students' favourite type of music, you could make a circle graph like **Figure 4**. In a circle graph, each piece stands for a different category (in this case, the kind of music preferred), and the size of the piece tells the percentage of the total that belongs in the category (in this case, the percentage of students who prefer a particular kind of music).

Figure 4

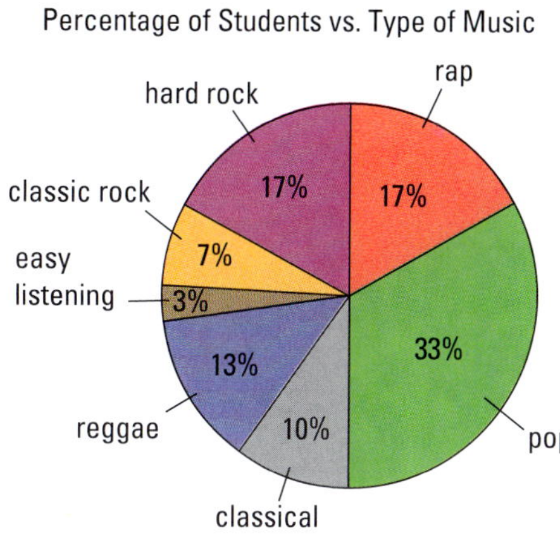

Try This Choose a Graph

Copy and complete **Table 3**. Determine the type of graph (point-and-line, bar, or circle) that is most suitable to illustrate the relationship between the variables in each pair.

Explain each of your choices.

Table 3

Pairs of variables	Most appropriate type of graph
1. Hardness of water Different cities/towns in Ontario	?
2. Amount of water used in a regular showerhead Amount of water used in a water-saver showerhead	?
3. List of five most abundant elements in the Earth's crust % abundance of elements	?
4. Surface area of an airplane wing Time the airplane stays aloft	?

7B Reading a Graph

When data from an investigation are plotted on an appropriate graph, patterns and relationships become much easier to see and interpret—it is easier to tell if the data supports your hypothesis or matches your design criteria. Looking at the data in a graph may lead you to a new hypothesis or to alter your design.

The graph in **Figure 1**, for example, clearly shows the relationship between the mass of an aircraft and the length of runway required for the aircraft to take off.

Figure 1 Aircraft mass vs. Length of runway required

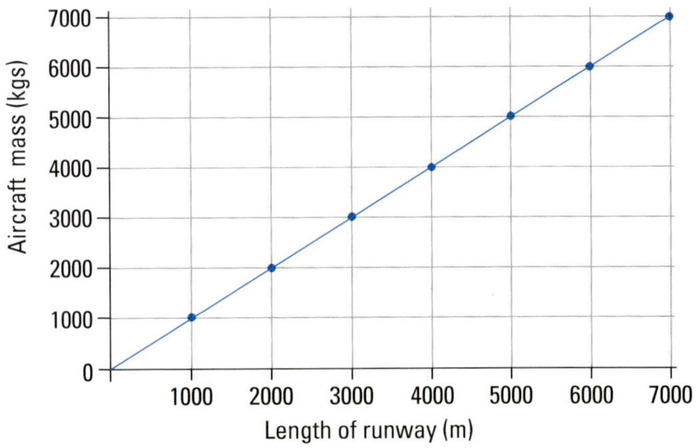

What to Look for When Reading Graphs

Here are some guiding questions as you interpret the data on a graph:

- What variables are represented?
- Is there a dependent variable? Is there an independent variable? If so, identify each.
- Are the variables quantitative or qualitative?
- If the data are quantitative, what are the units of measurement?
- Are two or more sample groups needed? If so, are they included?
- What do the highest and lowest values represent on the graph?
- What is the range (the difference between the highest and lowest values) of values for each axis?
- What patterns, trends, functions, or relationships exist between the variables?
- If there is a linear (straight-line) relationship, what might the slope (steepness) of the line tell us and what is the impact?
- Is this the best graph for the data?

Try This: Monitoring Water Quality

What does the graph in **Figure 2** tell you?

Figure 2 Phosphorous Levels in Lake Water

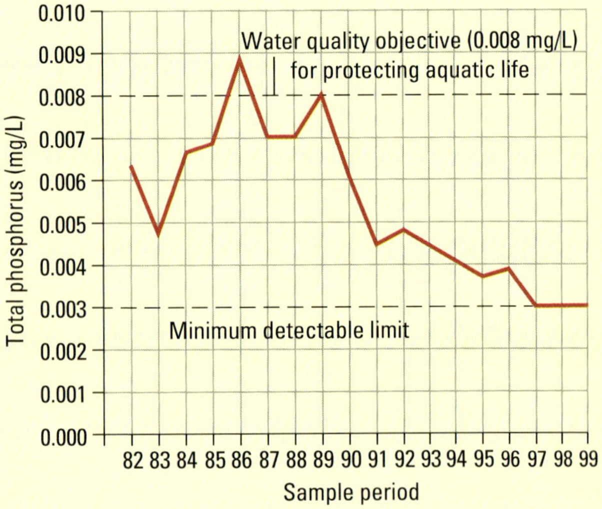

7C Constructing Graphs

Making Point-and-Line Graphs

Point-and-line graphs are common in mathematics, economics, geography, science, technology, and many other subjects. This section will help you become more skilled at drawing them, and at understanding point-and-line graphs produced by others.

As an example, the data in **Table 1** are used to produce a graph.

When making a point-and-line graph, follow these steps:

Table 1

Elevation (m)	Average height of tree (m)
1000	4
900	6
800	7
700	9
600	11
500	14
400	16
300	18
250	20
200	24
150	24
100	26
50	27
0	28

1. Construct your graph on a grid. The horizontal edge on the bottom of this grid is called the *x*-axis and the vertical edge on the left is called the *y*-axis (**Figure 1**). Don't be too thrifty with graph paper—if you draw your graphs large, they will be easier to interpret.

Figure 1

Step 1: Draw the axes.

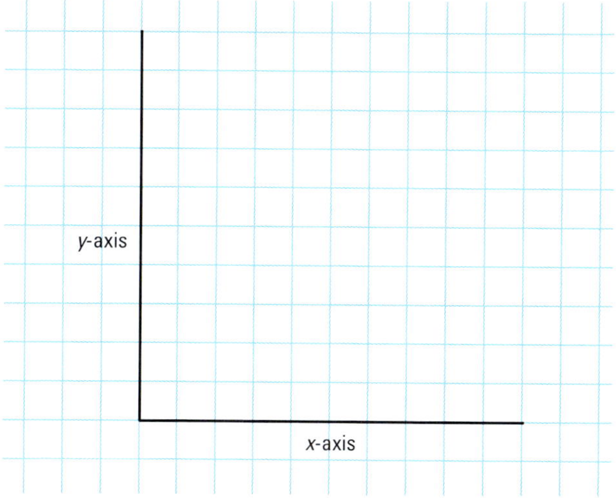

2. Decide which variable goes on which axis, and label each axis, including the units of measurement (**Figure 2**). It is common to plot the dependent variable (average height of tree in m) on the *y*-axis and the independent variable (elevation in m) on the *x*-axis.

Figure 2

Step 2: Label each axis.

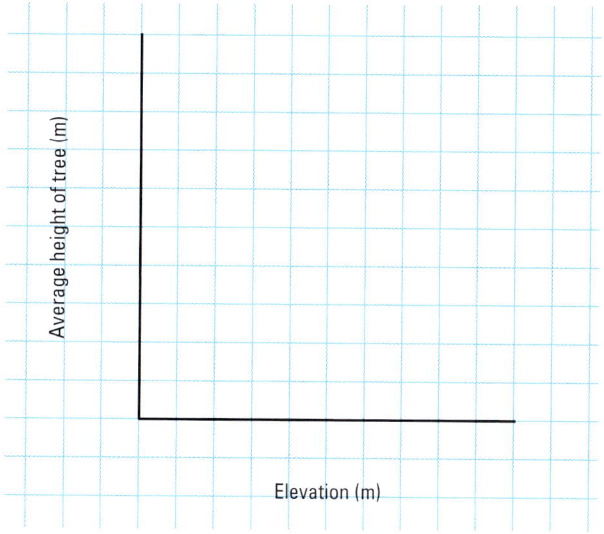

3. Determine the range of values for each variable. The range is the difference between the largest and smallest values. For the average height of tree, the maximum is 28 m, and the minimum is 4 m, so the range is: 28 m – 4 m = 24 m. For the elevation, the range is 1000 m – 0 m = 1000 m.

4. Choose a scale for each axis (**Figure 3**). This will depend on how much space you have and the range of values for each axis. Each line on the grid usually increases steadily in value by a convenient number, such as 1, 2, 5, 10, 20, 50, 100, etc. In the example, there are 10 lines for each axis. To calculate the increase in value for each line on the *x*-axis, divide the range by the number of lines:

$$\frac{28 \text{ m}}{10 \text{ lines}} = 2.8 \text{ m/line, which is rounded up to 3.}$$

Then, round up to the nearest convenient number, which in this case is 3. The scale on the "Average height of tree" axis should increase by 3 every space. Repeat the calculation for the *y*-axis:

$$\frac{1000\text{ m}}{10\text{ lines}} = 100\text{ m/line}$$

Figure 3

Step 4: Choose a scale for each axis.

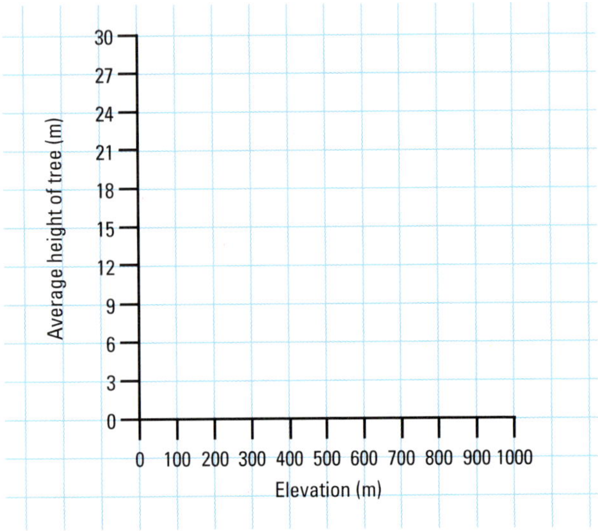

5. Plot the points (**Figure 4**). Start with the first pair of values from the data table, 1000 m elevation and 4 m trees. Place the point where an imaginary line starting at 1000 on the *x*-axis meets an imaginary line starting at 4 on the *y*-axis.

Figure 4

Step 5: Start plotting points.

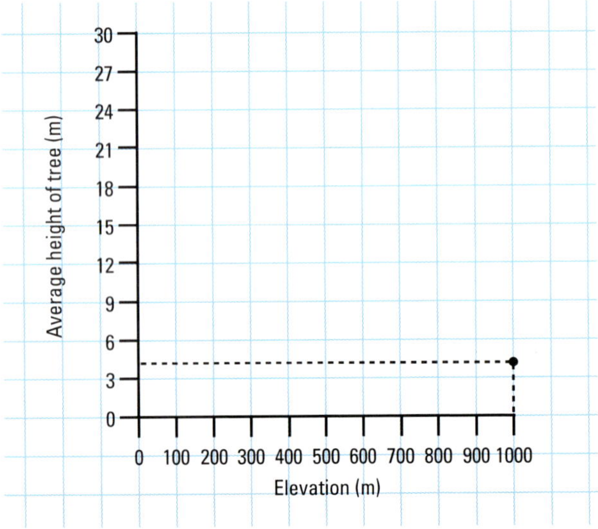

6. After all the points are plotted, and if it is possible, draw a line through the points to show the relationship between the variables (**Figure 5**). It is unusual for all the points to lie exactly on a line. Small errors in each measurement tend to move the points slightly away from the perfect line. You must draw a line that comes closest to most of the points. This is called the line of best fit—a smooth line that passes through or between the points so that there are about the same number of points on each side of the line. The line of best fit may be a straight line or a curved line.

Figure 5

Step 6: Draw a line of best fit.

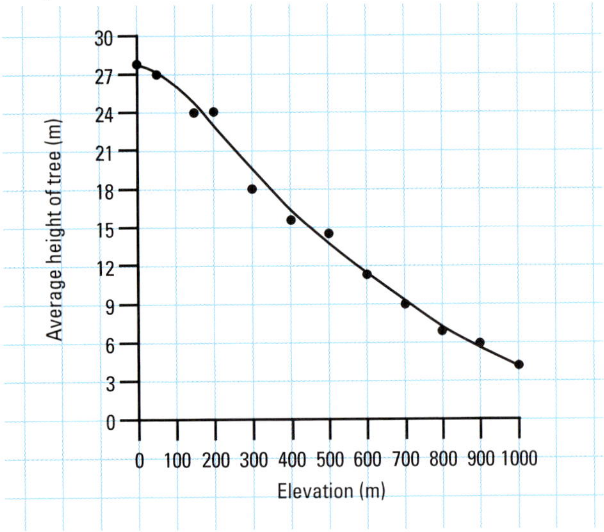

7. Title your graph.

Making Bar Graphs

Bar graphs are useful when working with qualitative data and when a variable is divided into categories. In the following example science students did a study of the kind of music students listen to while doing mathematics problems, and got the results listed in **Table 2**.

Table 2

Type of music	Math score (%)
easy listening	69
hard rock	60
jazz	72
blues	70
classical	80
folk	68
soft rock	75

Follow these steps to plot a bar graph of this data.

1. Draw and label the axes of your graph, including units (**Figure 6**). Some people prefer to have the bars based on the *x*-axis; others prefer to use the *y*-axis as the base. In the illustrations, the *x*-axis was chosen for the base.

Figure 6

Step 1: Draw the axes.

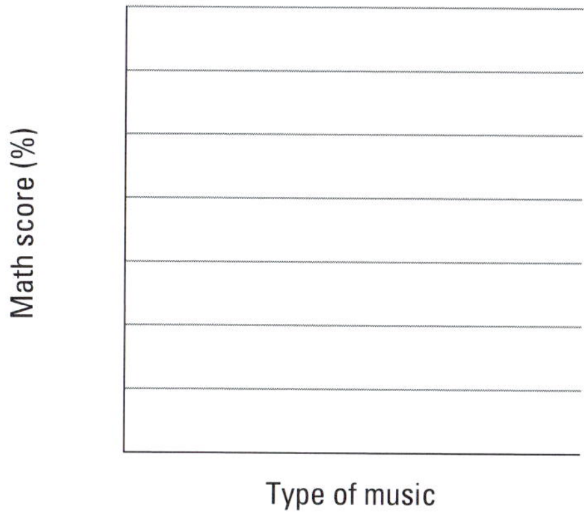

2. Develop a scale for the axis of the quantitative variable, just as you would for a point-and-line graph. In this example, the *y*-axis increases by fives, starting below the lowest value. In the illustration, 50 was chosen as the starting point, so all the bars would be visible.
3. Decide how wide the bars will be and how much space you will put between them. This decision is based on:
   - How much space you have. Measure the length of the axis on which the bars will be based, and divide that length by the number of bars. This will give you the maximum width of each bar.
   - How you want the graph to look. Decide how much less than the maximum width your bars will be, based on the visual appeal of thick and thin bars.
4. Draw in bars (**Figure 7**). Start by marking the width of each bar on the base axis. Then, draw in the top of each bar, according to your data table, and the sides. You can shade the bars equally or make each bar different from the others. It is important, however, to keep the graph simple and clear.

Figure 7

Step 4: Draw the bars.

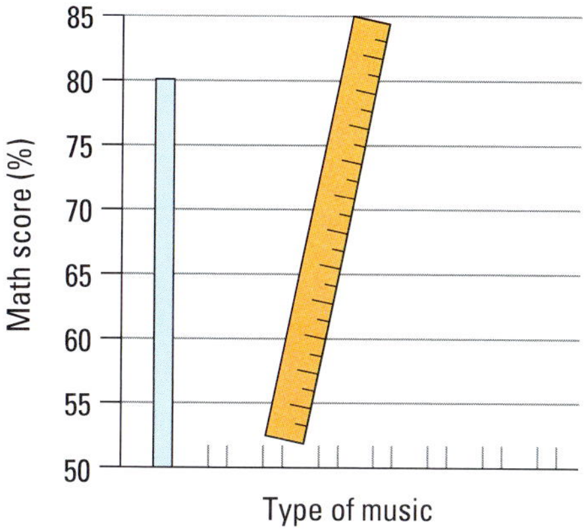

5. Identify each bar (**Figure 8**). There are several ways to do this. The best choice is the one that makes the graph easy to understand.

Figure 8

The completed bar graph

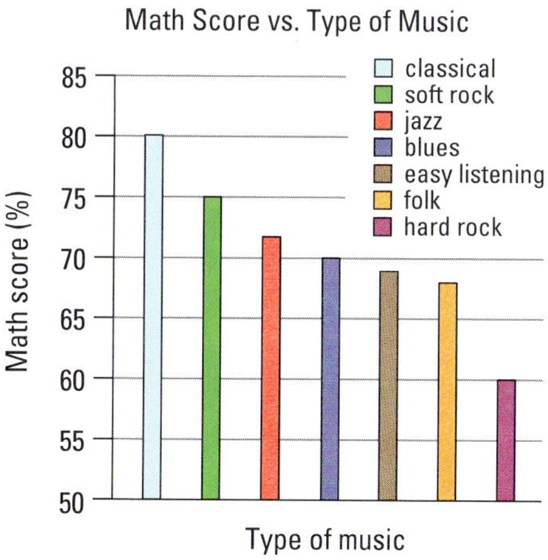

Making Circle (or Pie) Graphs

If your quantitative variable can be changed to a percentage of a total quantity, then a circle graph is useful. A sample circle graph is worked out below, using the data in **Table 3**.

Table 3

Type of music	Number of students who prefer that type	Percentage of total (% and decimal)	Angle of piece of pie (degrees)
rap	5	17% = 0.17	61.2
pop rock	10	33% = 0.33	118.8
classical	3	10% = 0.10	36.0
reggae	4	13% = 0.13	46.8
easy listening	1	3% = 0.03	10.8
classic rock	2	7% = 0.07	25.2
hard rock	5	17% = 0.17	61.2
TOTAL	30	100% = 1.00	360

Follow these steps to construct a circle graph.

1. Convert the values of your quantitative variable into percentages, and then into decimal form. In the sample, each number of students who prefer a type of music was turned into a percentage of the total number of students.

$$\text{Percentage} = \frac{\text{number}}{\text{total}} \times 100\%$$

$$\text{Percentage for rap} = \frac{5}{30} \times 100\% = 17\% \text{ (decimal version} = 0.17\text{)}$$

2. Multiply the decimal version of each percentage by 360 (there are 360° in a circle) to get the angle of each "piece of the pie" within the circle.

$$\text{Angle of piece of pie for rap} = 0.17 \times 360° = 61.2°$$

3. Draw a circle using a compass. To make the graph easy to read (and make), the circle should be big. The more pieces there are, the bigger the circle should be.

4. Draw in each piece of pie, using a protractor (**Figure 9**).

Figure 9

Step 4: Draw the pieces.

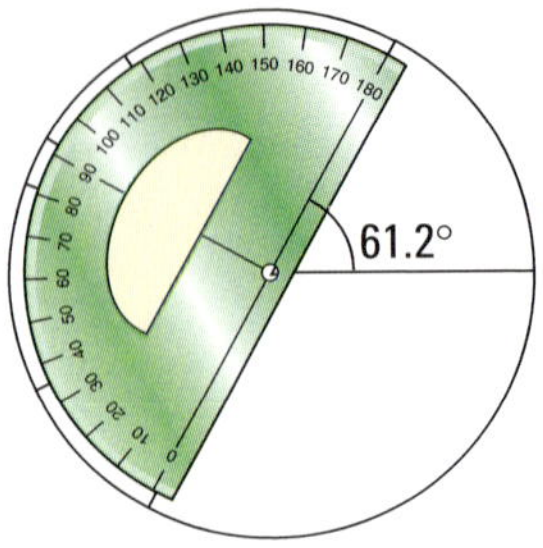

5. Shade each piece of pie using colours or patterns (**Figure 10**).

Figure 10

Step 5: Shade the pieces.

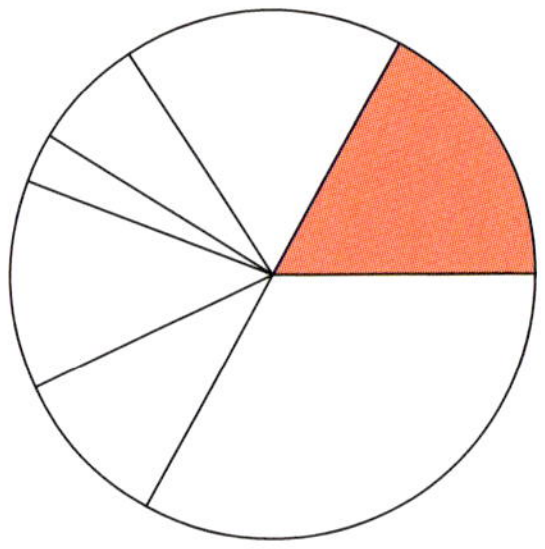

6. Label and title the graph (**Figure 11**). Put the percentage and the name of each category with its piece of pie (perhaps percentage inside and category outside the circle), or include them in a legend. Pick a title for your graph that describes the variables.

Figure 11

Step 6: The completed circle graph

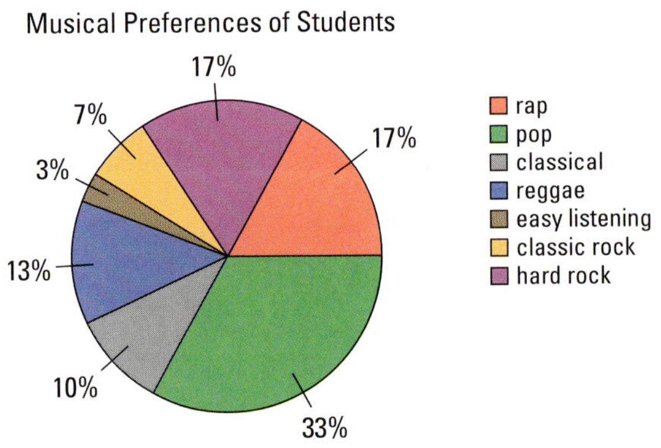

Using Computers for Graphing

You should be aware that there are many useful computer programs that can help with the graphing process. For example, Microsoft Excel is a very powerful spreadsheet/graphing program that allows for the construction of point-and-line, bar, and circle graphs as well as many other types of graphs (**Figure 12**). In addition, such programs can make use of statistical analysis to compute the best straight line or line of best fit.

Figure 12

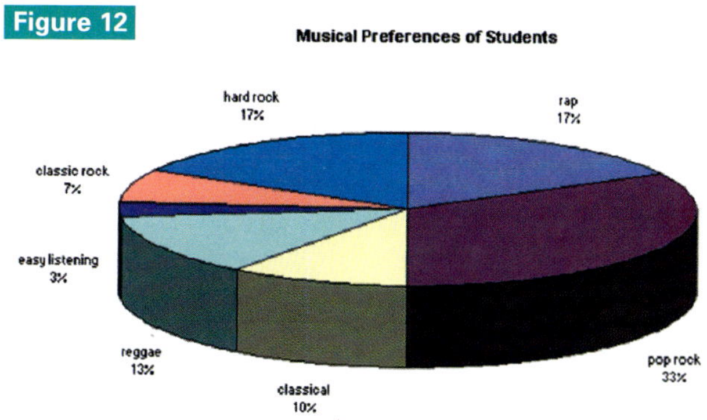

Try This Graphing

Perform the following tasks and graph the data that is given or that you collect.

1. Make labelled point-and-line graphs, with appropriate lines of best fit, for the data in **Table 4**.
2. Conduct a survey of your class or a larger group, and make bar and circle graphs from the data you collect. You could use one of the following variables or one of your own: most popular brand of sports shoe, favourite band, most interesting technological job, most well-known international structure or building.

Table 4 Amount of force needed to close a door vs. the position on the door where the force is applied

Distance from the fulcrum (hinges) in centimetres	Force applied in newtons
10 cm	80 N
15 cm	75 N
20 cm	70 N
25 cm	65 N
30 cm	60 N

7D Using Math in Science & Technology

Our world can often be analyzed, described, and predicted mathematically. For example, the mass and volume of a pure substance, when viewed in a point-and-line graph, prove to be directly proportional to one another. That is to say, as the volume of the object increases the mass of the object increases as well. If we set up a ratio of

$$\frac{\text{the mass of an object (m)}}{\text{the volume of an object (V)}} = \text{density of an object (D)}$$

we can define a new property—density—for a substance. This is just one example of how a mathematical relationship allows for the extension of our understanding.

While mathematics can sometimes appear abstract and daunting, as a scientist or technologist your ability to understand some fundamental mathematical concepts will improve your analytical ability.

Scientific Notation

In your studies, you use some very large and some very small numbers. For example, the average distance from the centre of the Earth to its crust is 6 400 000 m (**Figure 1**) and the average radius of a hydrogen atom is 0.000 000 000 05 m (**Figure 2**).

Because we don't want to spend the better part of our day writing out zeros, it is convenient to use a mathematical abbreviation known as scientific notation. The scientific notations for these two values are 6.40×10^6 m and 5.0×10^{-11} m, respectively. Can you see how this works?

Essentially,

multiply by the number 10 a total of 6 times

$$6.4 \times 10^6 = 6.4 \times 10 \times 10 \times 10 \times 10 \times 10 \times 10$$

$$5.0 \times 10^{-11} = \frac{5.0}{(10 \times 10 \times 10 \times 10 \times 10 \times 10 \times 10 \times 10 \times 10 \times 10 \times 10)}$$

divide by the number 10 a total of 11 times

Figure 1

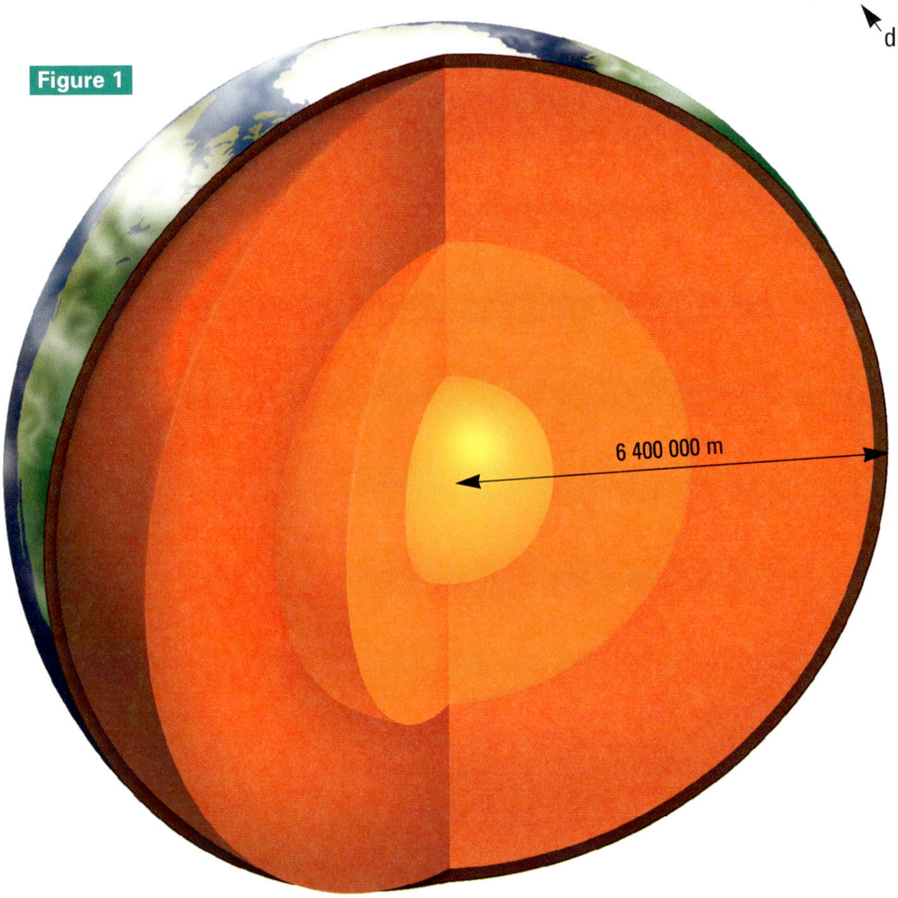

Figure 2

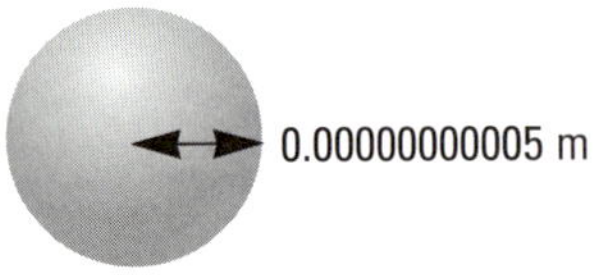

hydrogen atom

Significant Figures

Imagine you are given two pieces of string and you wish to find the total length of string that you have (**Figure 3**). You measure the length of one piece of string with an accurate scale ruler and find that it is 12.72 cm long. You measure the length of the other piece of string with an old metre stick and find that piece of string to be 14 cm long. What is the total length of string that you have? If you claim the total length to be 26.72 cm, you have assumed that the second piece of string is 14 cm—was your old metre stick able to give you this level of precision? No. You can only come to the conclusion that you have a total length of 27 cm.

Figure 3

When we take measurements and manipulate numbers in science and technology, we must look at the number of significant digits (or figures) in a number. The number of significant digits represents how carefully, and with what level of accuracy or precision, the measurement was taken.

Calculating Averages

There are many statistical methods that help us analyze the quantitative data that we collect. One of the most important of these is finding the average of a set of numbers. Calculating the average of a set of numbers allows you to reduce your findings to one representative value. For example, if we measure the heights of four students,

Suzanne = 175 cm	Jan = 185 cm
Omar = 145 cm	Molly = 180 cm

we could calculate that the average height of the group is

$$\text{average height} = \frac{175 + 185 + 145 + 180}{4} = 171 \text{ cm}$$

Predicting Using Formulas

Using various mathematical or experimental techniques, we can connect and explain phenomena by a mathematical formula. For example, you will study the following relationship:

$$\text{pressure} = \frac{\text{force}}{\text{area}}$$

If you know this formula, and you know any two of the quantities, you can predict the third. For instance, consider two people on a toboggan. If you know the area of a toboggan (1 m^2) and the force exerted on it by the two people sitting on it (1000 N), you can predict the pressure:

$$\frac{1000 \text{ N}}{1 \text{ m}^2} = 1000 \text{ N/m}^2 \quad \text{also known as 1 kilopascal (kPa)}$$

Try This — What Pressure?

A pile of scrap metal with a force of 20 000 N is dumped on a platform with an area of 25 m^2. What is the pressure on the platform?

7E Reaching a Conclusion

Scientific investigations mean very little, if anything, unless the scientist states a conclusion about the results. A conclusion is a statement that explains the results of an investigation. This statement should refer back to your original hypothesis. It should reveal whether the results support, partially support, or reject your hypothesis. Don't worry if your hypothesis is incorrect—scientists usually need to revise and repeat experiments many times in order to obtain the solutions they are seeking. Remember, science is a repetitive process. How many experiments do you think have been repeated in order to learn what we now know about cancer treatments?

In the process of design, the evaluation of the prototype is the stage at which conclusions are reached (see 3F "Testing and Evaluating a Prototype"). Your evaluation will take into consideration how effectively you solved the initial design problem. The design process is also a repetitive process: products and processes are tested and modified and, ultimately, improved.

Figure 1

Consider all the variables when you reach your conclusion.

Checking Your Hypothesis

Suppose you wanted to find out about the relationship between the amount of time spent doing homework per week and the mark you receive on your tests. Your hypothesis may be "As I increase the amount of time I spend on Science & Technology homework, my level of success on tests increases." Over a period of 10 weeks, you increase the number of hours you study by 2.5 h per week.

At the end of each week, you are given a test and record your results. At the end of 10 weeks your results have improved.

One conclusion that could be made from these results might read, "My hypothesis was correct. The results show that when I spend more time on Science & Technology homework, there is an increase in my test scores because I understand the material better." There may be other possible hypotheses and conclusions that would explain these results. When reaching a conclusion, it is important to consider all the variables, e.g., your study methods, the environment in which you studied, and the time of day (**Figure 1**).

Reaching conclusions in Science & Technology investigations allows you to critically analyze results using a mix of logic, common sense, understanding, and patience.

Try This — Think Again

Consider the variables that you would have had to control in order to conduct a sound investigation to test the effect of hours of study on test results. What possible hypotheses might you want to consider before settling on your final conclusion?

7F Reflecting on Your Work

It is always important to reflect on events in order to learn from them. This is one aspect of inquiry in science and technology that is sometimes neglected, especially with beginners. Once you have been through either the process of inquiry or design, it is always advantageous to step back and think about what you did. What went well? What were the challenges? What could be improved? What would you do differently if you were to go through the process again? It is through reflection that you will improve your work and increase your learning.

When reflecting back on the process of inquiry or design, be sure to identify the types of error that may have emerged. These sources of error should be identified in the conclusion or discussion of your results and be used as the basis for improving the process the next time. For example, let's say you completed an experiment and were able to conclude that plants did grow toward sunlight. What sources of experimental error could have occurred while performing this experiment? These could include such things as variations in air temperature, inconsistent exposure to light, or differences in soil composition. Be as specific as possible when stating experimental errors. It is not good enough to simply state: "It was due to human error." Reflection is equally important when evaluating the design process. For example, once you've developed the prototype for a new type of school chair, you need to think about what modifications or improvements you should make.

In the process of scientific inquiry, experimental errors decrease the validity of an experiment. But, more importantly, they allow you to revise your procedure. That revision will increase the reliability of the results the next time you perform the same investigation. Remember, experiments are a repeating process. By repeating the process, you are improving your ideas and investigative skills.

In any investigation, it is important to recognize that what may appear to be an error in one phase may ultimately turn out to be the means to improvement or innovation.

Try This **If You Did It All Over...**

Complete one of the tasks below:

- Think back to an argument you had with your parents or a friend. Now, reflect on the conflict from beginning to end. List all the things that you would do differently if you were ever in that situation again.
- Think back to a scientific experiment or technology project you have done in the past. Now, reflect on the work you did from beginning to end. List all the things that you would do differently to make it more reliable.
- Think back to a prototype you have built in the past and identify the modifications you would make in order to improve it.

8A Writing a Report

All investigators use a similar format to write reports, although the headings and order may vary slightly. Indeed, your report should reflect the process of scientific inquiry that you used in the investigation.

The Features of a Report

Cover Page: Make a cover page (**Figure 1**) that includes the following:

- the title of your investigation
- your name
- name of your partner(s) (if applicable)
- your instructor's name
- the due date

Figure 1

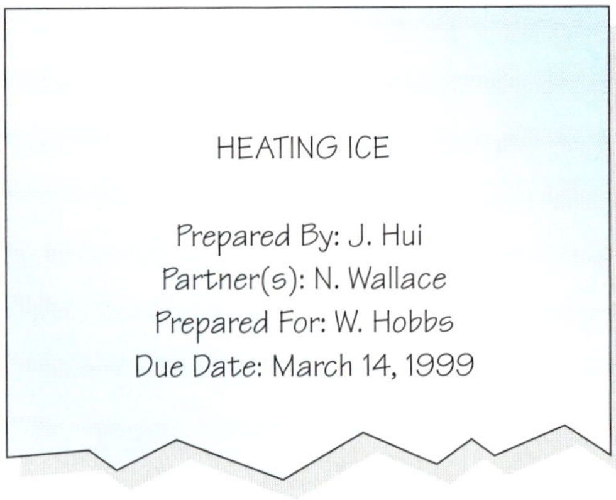

HEATING ICE

Prepared By: J. Hui
Partner(s): N. Wallace
Prepared For: W. Hobbs
Due Date: March 14, 1999

Title: At the beginning of your written report, write the title of your experiment.

Introduction: Always begin with a brief explanation of pertinent theory underlying the experiment. This includes the information you discovered by researching your topic. This section is also referred to as your "review of literature."

Purpose, Question, or Problem: Make a brief statement about your investigation. This can be written as a question.

Hypothesis: Write the hypothesis. Remember this is your "educated guess," based on your previous knowledge and the research you completed.

Materials: These include consumables (e.g., water, paper towels). Be specific about sizes and quantities. These also include non-consumables (e.g., test tube, beaker). Make a detailed list of the materials you used.

Diagram: Make a full-page diagram of the materials and equipment you used in the experiment (**Figure 2**). Remember to label and title the diagram. Your diagram can be placed at the end of your report, following the applications.

Figure 2

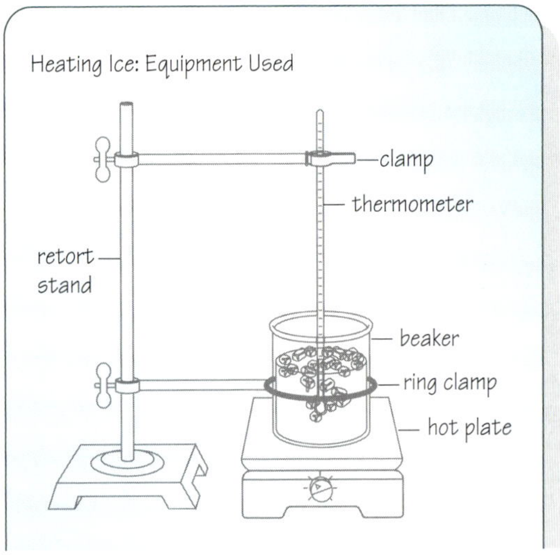

Experimental Design: List the independent, dependent and controlled variables in your experiment and summarize the procedure.

Procedure: The most important part of an investigation, when others are trying to determine if it is "good" or "bad" science, is the procedure. To be sure that your work is judged fairly, make sure you leave nothing out! Remember to write this in numbered steps, past tense, and passive voice.

Observations/Results: Present your observations and results in a form that is easily understood. The data should be in tables, graphs, or illustrations, each with a title. Include any calculations that are used. The results of the calculations can be shown in a table.

Analysis/Conclusion: Summarize the investigation as you would if you were writing a book report. Refer back to your hypothesis. Was it correct, partially correct, or incorrect? Explain how you arrived at your conclusion(s). Justify your method and describe your results. Suggest a theory to support or interpret your results. If you were assigned questions with an investigation, you would answer them here. Discuss any sources of experimental error that may have affected your findings.

Applications: Describe how the new knowledge you gained from doing the investigation relates to your life and our society. It should answer the question, "Who cares?"

References/Bibliography: Give credit for the resources you used in your research. Always cite your source(s). Failing to do so is considered plagiarism (unacknowledged copying). It is unethical and illegal.

Citing Your Information Sources

Giving Credit for Material Used Within the Report

Whenever you give credit to an author (including yourself from previous reports!) for the use of graphs, tables, diagrams, or ideas, use the following technique:

Immediately after the information is used, give the last name of the source, the date of the publication, and the page reference. For example:

> The bedrock may be under soil, but it is still subject to biological weathering from plant roots (Gibb, 1999, p.22).

Writing a Bibliography

The bibliography appears at the end of your report and includes the full reference for every source you cited within your report as well as any sources you used for general reference, fact-checking, etc. It should be in alphabetical order of authors' last names. Be sure to use hanging paragraphs. The format for books, journals, web sites, newspapers, and CD-ROMs differ. To make sure you cite your sources correctly, refer to the following examples:

- **Information from a book:**

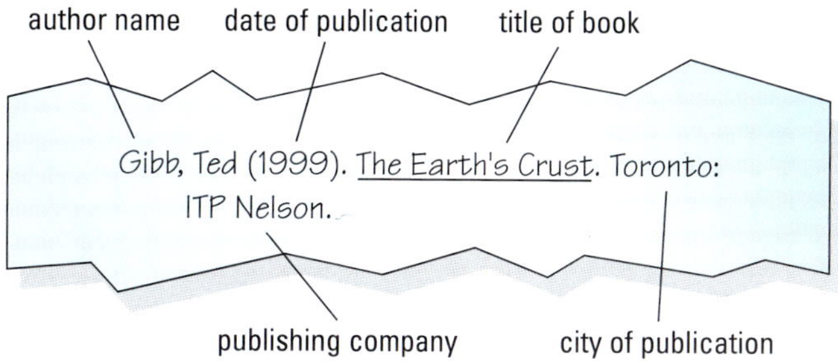

- **Information from a journal or a magazine:**

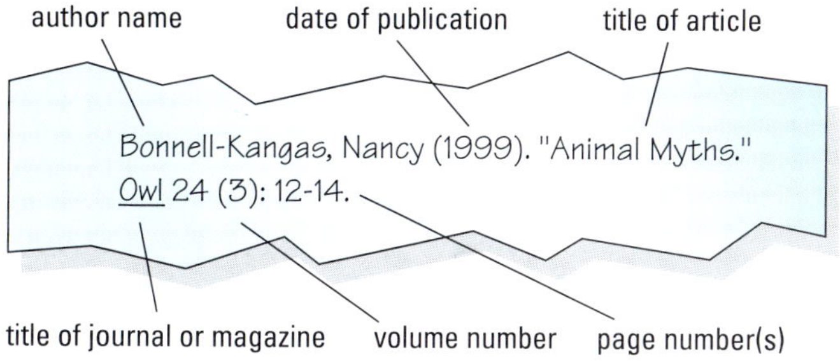

- **Information from a newspaper:**

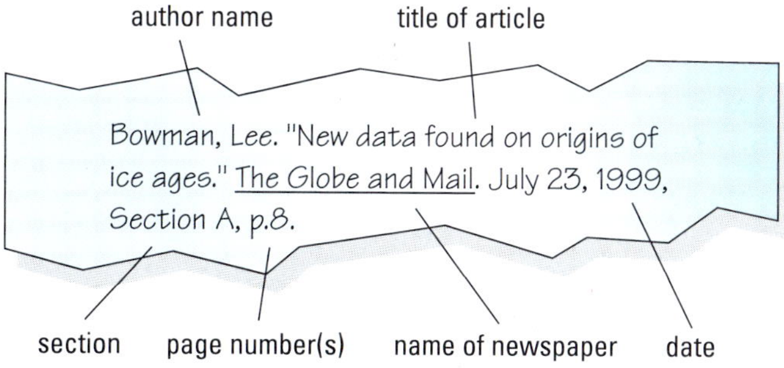

- **Information from a web site:**

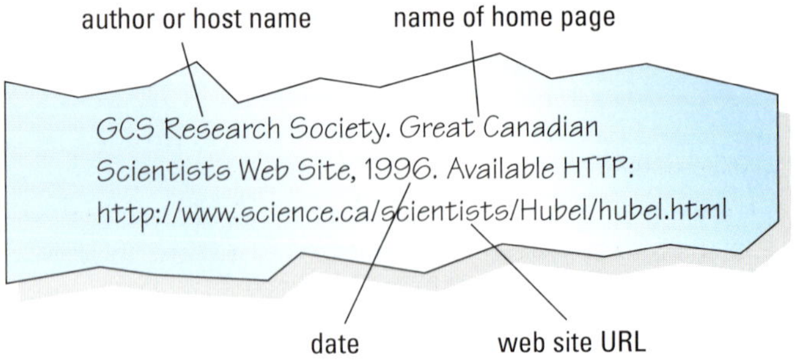

- **Information from a CD-ROM:**

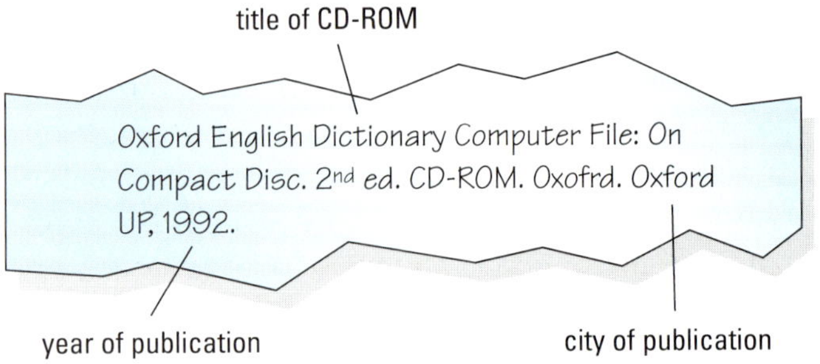

Checklist for Writing a Report

Refer to this checklist whenever you are required to write a scientific report.

✓ Cover page
✓ Title of the investigation
✓ Introduction
✓ Purpose/Question
✓ Hypothesis
✓ Materials
✓ Diagram
✓ Experimental Design
✓ Procedure
✓ Observations/Results
✓ Analysis /Conclusion
✓ Application
✓ References/Bibliography

8B Creating a Design Folder

It is a good idea to create a folder that communicates important information about your design. The end product (or process) will rarely speak for itself. You should be able to explain why your design is useful, what steps you took in the design process, and how well the prototype meets the needs for which it is designed. In fact, you should summarize all the steps you took in the process of design as outlined in 3A "The Problem-Solving Cycle."

Features of a Design Folder

Cover Page: Create a cover page (**Figure 1**) that includes:

- name of the product or process you designed
- your name
- your partner(s) name (if applicable)
- the date the folder was completed

Figure 1

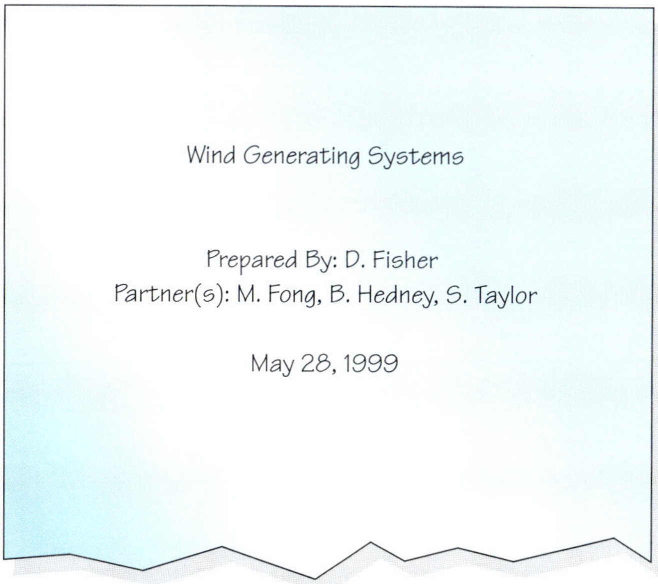
Wind Generating Systems

Prepared By: D. Fisher
Partner(s): M. Fong, B. Hedney, S. Taylor

May 28, 1999

Title: At the beginning of your folder, write the name of your product or process.

Problem: State your problem briefly and clearly and define any key terms.

Introduction: Provide background information on your problem. What was the context for your design? How did you assess the need that your design is intended to satisfy? If you conducted any interviews or surveys, summarize the results here. (Details can be placed in an appendix.) Summarize any research you did on existing products or processes that might meet your need, and explain if and how these were adapted to solve the problem you identified.

The Best Alternative: Explain how you chose the solution you did. List all the constraints, and explain and rate the design criteria. Itemize all the alternatives that you considered as possible solutions to the problem, and explain how you rated them. Include any useful charts or graphs. For instance, if you used the problem-solving matrix to choose the best alternative, include the results in this section (**Figure 2**).

Figure 2

Rating Scale: 0=Bad

Alternatives	Specifications					Total
	Safety	Cost	Durability			
	0-10	0-9	0-8			
Solar Paneled Ceilings	6	2	3			11
Water Wheel	4	0	2			6
Windmills	8	6	6			20

Design Plan: Explain how your prototype was put together and how it is supposed to function. Include well-labelled sketches and drawings to support your explanation (see 6C "Science & Technical Drawing" for samples.) Include a list of necessary materials and tools. If appropriate, include operating instructions.

Tests and Evaluation: Present the results of your prototype tests. Be sure that your tests measured the prototype against the specific design criteria you used. Use tables and graphs, and/or qualitative observations to illustrate your results. Write a brief paragraph evaluating how well the prototype met your design criteria—how well did it solve your problem?

Summary: From all of your experience and data, draw conclusions as appropriate. Be certain the conclusions are based on the data obtained from the construction of your prototype, the test results, and from a thorough literature survey. Describe what worked and what didn't. Indicate what could be improved if the prototype were rebuilt. Present evidence to support your conclusions.

References/Bibliography: In addition to citing various sources (see 8A "Writing a Report" for the correct format for citing sources of information), include a list of all the people who have made substantive contributions to your project.

Checklist for Creating a Design Folder

Refer to this checklist whenever you are required to create a design folder.

✓ Cover page
✓ Name of the product or process
✓ Introduction
✓ Problem
✓ The Best Alternative
✓ Design Plan
✓ Tests and Evaluation
✓ Summary
✓ References/Bibliography

8C Multimedia Presentations

Multimedia presentations provide a lot of scope for creativity. They allow you to use a combination of colour, computer graphics (even animation), sound and video clips to communicate information in an interesting, interactive way. When you read a book, for example, you tend to get information in a linear fashion, one page after another. When you interact with a multimedia presentation (e.g., web sites, computer games, CD-ROMs, kiosks) you have much more control over what information you see and when you see it.

There are a number of different software programs that can help you create effective multimedia presentations. There are specific skills you need to use specific programs. What you can create will depend on what hardware and software you have available to you, and how well you know how to use them (or how much time you have to learn!). There are some general tips you should bear in mind, no matter what kind of multimedia presentation you are creating.

General Tips

- Be sure to have a definite and well-articulated purpose. Be sure that you know who your audience will be.
- Start by considering your introduction (where you state your main goals) and your conclusion (where you restate your main goals).
- Create a flowchart for your presentation (**Figure 1**). Consider how your audience might want to make use of the information you are presenting, and provide logical links from one place to another.
- Consider the length of your presentation. Be clear and concise!

Figure 1

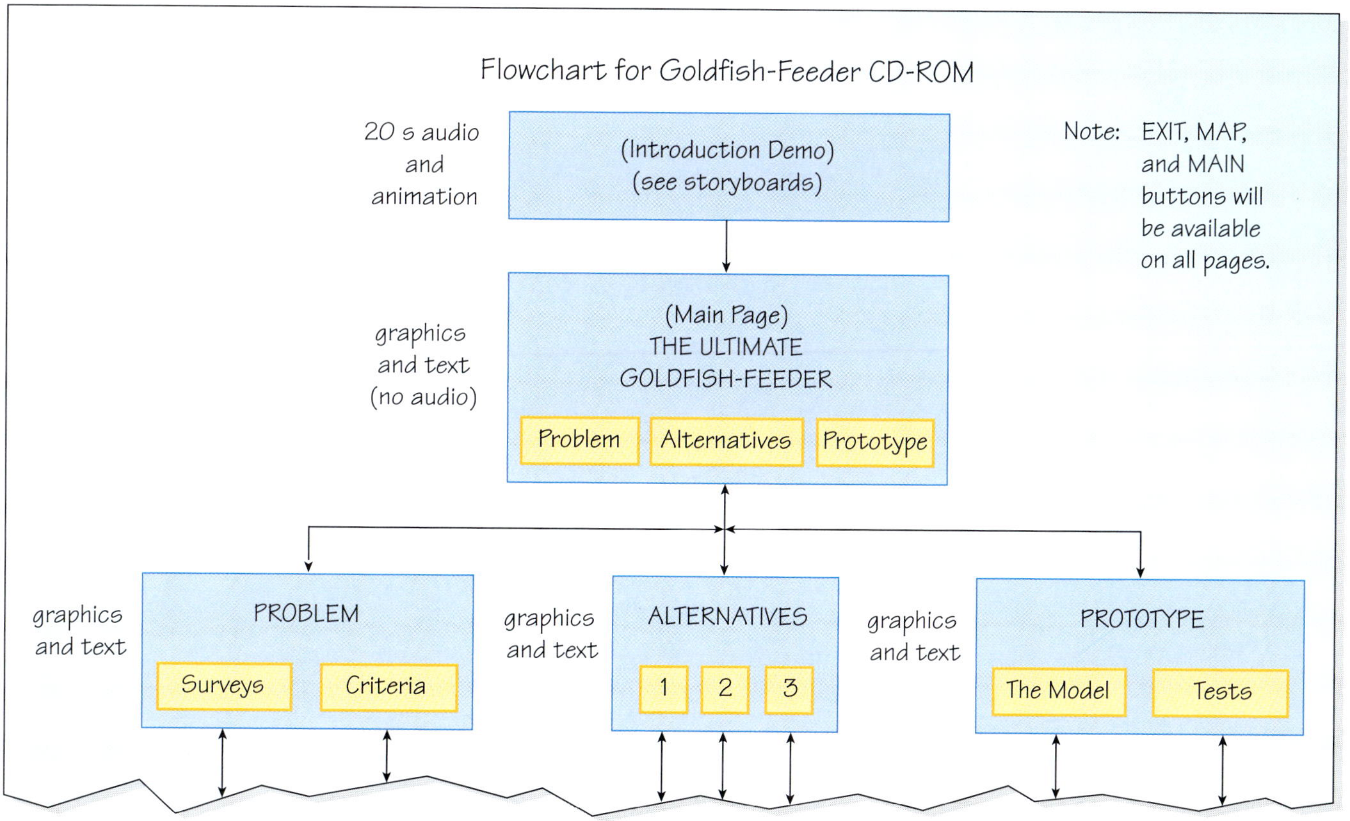

- Draw storyboards for your presentation (**Figure 2**). Be sure to consider all media: what are people going to see and hear at any given moment?
- Choose your audio-visual material carefully. Remember that the special effects you use should have a purpose. They should make your presentation more memorable and clarify any information you are presenting.

Try This — What Makes a Good Site?

Search the Internet for three good web sites. Choose one and analyze the way the information is presented. What are the strengths? List three ways you would improve the site. Write an e-mail to the webmaster stating what you like about the site and how it could be improved. The next time you access the site, some of your suggestions may have been implemented!

Figure 2

Frame #: 1	Length of shot: 2 s
	Visual: Black Audio: no sound

Frame #: 2	Length of shot: 3 s
	Visual: Fade-in on front-view of goldfish. Bubbles coming from mouth. Mouth opening and closing. Background becoming lighter—turning a clear blue. Audio: Bubbling sound of fish tank.

8D Exploring an Issue

In *Nelson Science & Technology 7/8*, you will have opportunities to explore many issues. Advances in technology and science need to be evaluated from many different perspectives—particularly since these advances are being made at an ever-increasing rate.

Figure 1 shows a helpful way of organizing the advantages and disadvantages of a given issue related to technology or science.

Figure 1
Should you use an electric toothbrush?

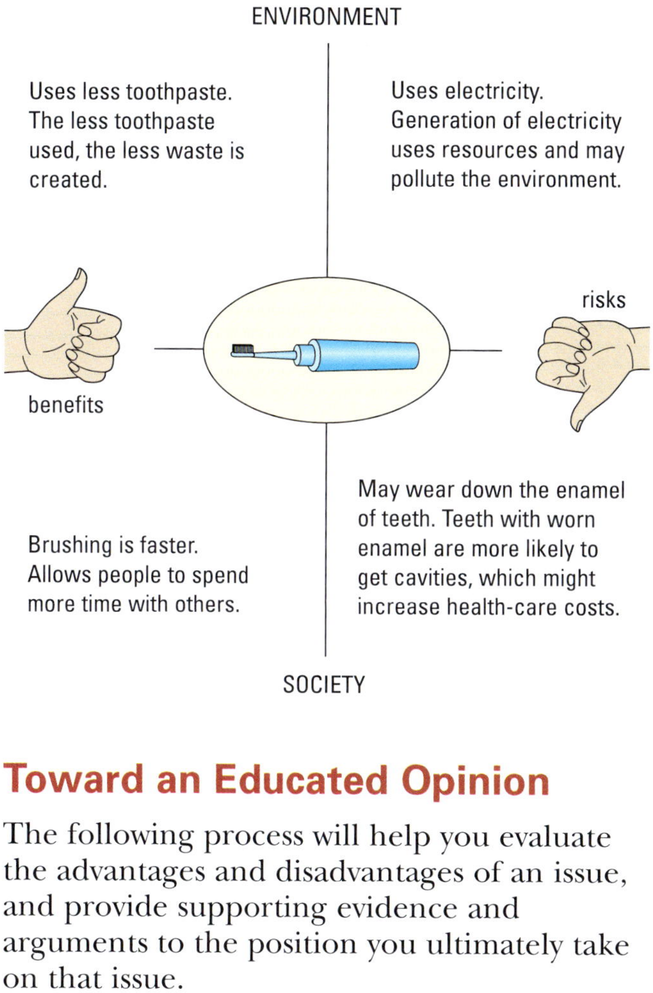

Toward an Educated Opinion

The following process will help you evaluate the advantages and disadvantages of an issue, and provide supporting evidence and arguments to the position you ultimately take on that issue.

1. Initial Research

Choose an issue that really interests you. Gather sources of information about it. Consider all the sources of information available to you (see 4A "Research Skills").

2. Formulation of a Question

Put together a question around the issue. The question should be answerable, important to our society, and debatable. You are encouraged to try this question out on your classmates, teacher, and others.

3. Further Research

Continue your research on this topic. Find at least two additional sources of information on the issue. Be sure to scrutinize the credibility of your information. Remember PERCS (see 4A "Critical Thinking")!

4. Reaching Your Position

Answer your question. Explain your position thoroughly. Be sure to consider the following:

- Have you included information from at least three articles?
- Have you stated your position clearly?
- Have you shown why this issue is relevant and important to our society?
- Have you included at least two solid arguments (with solid evidence) backing up your position?
- Have you included at least two arguments against your position and shown their respective faults?
- Have you analyzed the strong points and weak points of each perspective?

5. Communicating Your Position

There are several ways of communicating your position, each with its own format:

- Write a brief essay.
- Participate in a formal debate.
- Participate in a role-play activity, taking on the role of a specific, affected party.
- Write a letter to the editor of a newspaper.

9A Setting Goals and Monitoring Progress

Think back to the time you spent in school last year. What classes did you do really well in? Why do you think you were successful? What classes did you have some difficulty with? Why do you think you had those troubles? What could you do differently this year that would help you do better than you did last year?

What you have just done is reflect on your past behaviour in an attempt to make positive change. The things that you want to accomplish today, this week, this year—throughout your life, for that matter—are called goals. Learning to set goals and planning toward their achievement takes skill and practice.

Setting Goals

- **Honestly assess your strengths and weaknesses.** Setting goals starts with reflection. Maybe you've noticed that you do better on projects and reports than you do on tests. Your strengths may include teamwork, independent research, and analysis. Your weaknesses may include inattention in class, anxiety or stress during tests, and poor note-taking skills.
- **Brainstorm a list of realistic goals that are important to you.** Don't set yourself up to fail by setting goals that you can't achieve, or that you don't care about. State your goals in relation to your expectations of yourself—not in comparison with the performance of others. For instance, "I will increase my test results by 10% by the end of the semester" is a better goal than "I'll have the highest test results in the class by the end of the semester."
- **Make sure your goals are measurable.** You will find it easier to keep trying to reach your goals if you can tell whether or not you are getting any closer to them. Again, "I will increase my test results by 10% by the end of the semester" is a better goal than "I'm going to do a lot better on my tests."
- **Discuss your goals with someone you trust.** Other people can often help you clarify your goals. People who know your strengths and weaknesses may think of possibilities you haven't considered. Sharing your goals with someone else is also a good way of enlisting outside support.

Planning Toward Your Goals

Once you have a list of realistic goals, you need to create a plan to achieve them. There are two essential components to your plan: actions and target dates.

Taking Action

The first thing to do is to make a list of actions that might help you reach your goals (**Figure 1**). If you've made an honest assessment of your strengths and weaknesses, you know what you have to work on to achieve your goals. If note-taking is one of your weaknesses, you could strengthen that skill by arranging to meet regularly with someone who could help you develop it. If

Figure 1

Goal: To increase my test results by 10% by the end of the semester

Possible actions:
- call Mark and Akim. Ask them if they want to start a study group for tests
- meet with Mrs. Rankin to ask how I could learn to take better notes
- increase the amount of time I study each evening by 30 minutes (NOTE: this will mean giving up something else! Must decide what!)
- reorganize my desk and bookshelf
- buy a day-timer
- schedule my homework time each day
- take Fridays off if I keep my schedule from Monday to Thursday

working in groups is one of your strengths, you could build on it by starting a study group to help you prepare for tests. The key is to identify what is preventing you from achieving your goal and to plan ways of overcoming those obstacles.

When you are making your list, include incentives and rewards. If overcoming obstacles starts feeling like a punishment, you won't work at it as long or as hard.

Making a Schedule

Your goal is a target. You need to know what you want to achieve, and you need to know how long it should take to achieve it. Once you've made a list of actions, you can assign dates to them. These dates provide short term targets that, if you hit them, will make it easier to hit your overall target.

Work backwards from your ultimate target date. For instance, if you hope to achieve a 10% increase in test results by the end of the semester, how much time have you got? How many tests are scheduled between now and then? What can you achieve before the first test? What more can you achieve before the second test?

Go back to your list of actions and write in a date beside each one. Stay flexible. Your plan will evolve as you put it into action. Often, your plan relies on other people and will change depending on their schedule. A working schedule is represented in **Figure 2**.

Figure 2

Sept. 19:	Organize study space
Sept. 20:	Meet with Mrs. Rankin: 3:55 p.m.
	Buy day-timer
Sept. 22:	First study group (plan to meet every Wednesday)
Sept. 23:	Science & Technology Study: 4:15-5:15 (same time every Mon., Tues., Thurs.)
Sept. 24:	TEST: 11:00 a.m.
	Meet Sara at Mike's: 4:10 p.m. (if goals met during the week)

Once you've roughed out a schedule that looks manageable, transfer the actions from your list onto a calendar. Keep your calendar in a prominent place in your study area, and make it a habit to refer to it on a daily basis.

Monitoring Progress

When you reach your ultimate target date, you have to be able to assess whether or not you've achieved your goal. It is also important to measure your progress along the way. The more detailed and specific your plan is, the easier this will be. In fact, your plan should include regular monitoring. You might decide to review your progress at the end of every week, for instance. How many of the things that you scheduled did you actually do? For example, did you have any tests that week? If so, did your results show an increase?

The results of this monitoring may be a change in plan. For example, maybe the study group idea isn't working very well, and you're spending more time talking than studying. It's always possible to change your plan.

Don't be too hard on yourself when you review your progress. If you missed some small targets, put it behind you and keep moving forward. Adjust your goal if you have to, but don't give up.

Try This — Your Goals

On a piece of paper, write three personal goals that you would like to accomplish by the end of the term or semester. Identify ways in which you will try to reach each of these goals. State how you will celebrate when you reach your goals.

9B Good Study Habits

Understanding anything—whether it is a life-saving technique in the swimming pool, a trumpet solo, or a Science & Technology lesson—is an active process. Studying takes on many forms. It involves learning and understanding material. Developing the following study skills can help you in your learning. You can modify these tips to help you in other school courses and in recreational activities.

Your Study Space

- **Organize your working area.** The place where you study at home and at school should be tidy and organized. Papers, books, magazines, or pictures that are strewn all over your working area will distract you from focusing on the work at hand.
- **Maintain a quiet study space.** Make sure that the place where you study is removed from distractions such as the phone, stereo system, TV, friends, and brothers and sisters. Popular study spaces include the school library, the public library, and your bedroom at home (if it has a working area). Any quiet place where you can be productive will work.
- **Make sure you feel comfortable in your study space.** You will be the most productive and study effectively if you are working in an area where you feel at ease—personalize your space.
- **Be prepared with all the materials you will need.** It is important that you have all your notebooks, textbooks, computer equipment, paper, pencil, pen, ruler, dictionary, thesaurus, and anything else you use for your work in your study space. If you have to continually get up to find a book or eraser, you won't be able to accomplish as much as you had hoped.

Study Habits

- **Prepare for class by reading material ahead of time.** It is also helpful to read or view materials from other sources, such as science magazines, newspapers, the Internet, and television programs.
- **Take notes.** To make note-taking easier, you may want to make up a shorthand method of recording ideas.
- **Review any notes you made in class the same day and add comments.** Then have a friend or relative quiz you on the material in the notes. Reinforce your understanding by answering the questions in the textbook—even if they are not assigned.
- **Use your notes and the textbook to prepare summaries.** Studying is most effective with a pen or pencil or your word processor, so you can write down the important ideas (**Figure 1**). You may want to write or type study cards to assist in making effective, point-form summaries of your notes. Look at the example in **Figure 2**. Notice there is a title at the top and all the information has been condensed into a point-form, easy-to-learn format. It's much more effective than learning material that is in paragraph form. Condense, condense, and condense!

Figure 1

- **Use graphic organizers to help you summarize a unit, or lesson.** (see 9E "Graphic Organizers"). You may want to use unit summaries in the student resources.
- **"Practice makes perfect."** This is as true for Science & Technology as it is for playing piano and shooting baskets. If you practise your study skills until they become almost automatic, you will have more time to think about how you will use them.
- **Schedule your study time.** This will help you avoid that most ineffective of all study methods, "cramming" before assignments and tests. Use a daily planner and take it with you to every class. Write all homework assignments, tests, projects, and extra-curricular commitments in it. This will assist you in organizing a daily "To Do" list that will ensure maximum use of your time.
- **Know your strengths and your weaknesses.** Take advantage of all opportunities to get help with areas in which you may have trouble. Use your strengths to help yourself and others. Form a study group and have regular meetings. You may be able to help others in some parts of the course. In turn, you may receive help from them.
- **Teach the material you have learned to someone who has not yet learned it.** Their questions will help you see what areas of the subject you need to learn more about and what areas you don't completely understand (**Figure 3**).

Figure 2
A study card

The Visible Spectrum

White Light = Roy G. Biv
- red
- orange
- yellow
- green
- blue
- indigo
- violet

- **Take study breaks.** It is important that you set study goals and take a short study break after meeting each of the goals you set. For example, you could decide that you will take a study break after crossing two items off your "To Do" list. Taking study breaks will help you rejuvenate for the next tasks on your list and will assist you in completing all your work effectively and accurately.

Study Checklist

✓ Organize your study environment. (Is it quiet? Comfortable? Organized?)
✓ Read material ahead of time.
✓ Take notes.
✓ Review.
✓ Summarize. (Make study cards, if that technique works for you.)
✓ Use graphic organizers.
✓ Practise.
✓ Schedule study time. (Use your daily planner and make a "To Do" list.)
✓ Know your strengths and weaknesses.
✓ Teach what you have learned.

Try This — Making Study Cards

Condense the information in this section of the *Skills Handbook* (9B) onto study cards. Remember to put a title at the top of each study card and make point form notes.

Figure 3

9C Using Your Computer Effectively

Computers are becoming more and more common in your learning. Most students have access to this technology at some point every week. Like your notebook, it is important to keep your computer files organized. It is also necessary to use your computer effectively—there is a time for computer work and a time for computer games and recreational surfing of the Internet. The following hints can assist you in using your computer effectively in your learning. Remember these hints can be modified to suit your needs.

General Computer Hints

- **Create a folder for all your science and technology work.** It is important that any computer work you do is kept in one folder. Otherwise, your work could be saved in lots of different locations on your computer and this means that it could become easily misplaced or even lost.
- **Organize your science and technology units in your folder.** Now that you have created a science and technology folder, organize this folder into the various units that you will study this year. Refer to **Figure 1** for a detailed glimpse of what your folders might look like.
- **Think carefully about what to name each of your documents.** Make sure that the names you give will allow you to easily identify what a particular document contains. Work that is called *Science & Technology 1*, for instance, provides no indication of what the work actually is. Be as specific as possible.
- **Back up your work.** Always make an extra copy of any work you do on the computer. Copy your work onto a disk or onto your school network. Too often students lose their work because they do not take the time to do this. It's much easier than having to complete a report or assignment all over again.
- **Maintain correct posture and form.** It is important that you don't sit too close to the screen and that you sit in an upright position with your hands formed correctly while typing. Incorrect hand positioning can eventually cause carpal tunnel syndrome. The study of how people interact effectively and safely with computers is a part of what is termed "ergonomics."

Figure 1

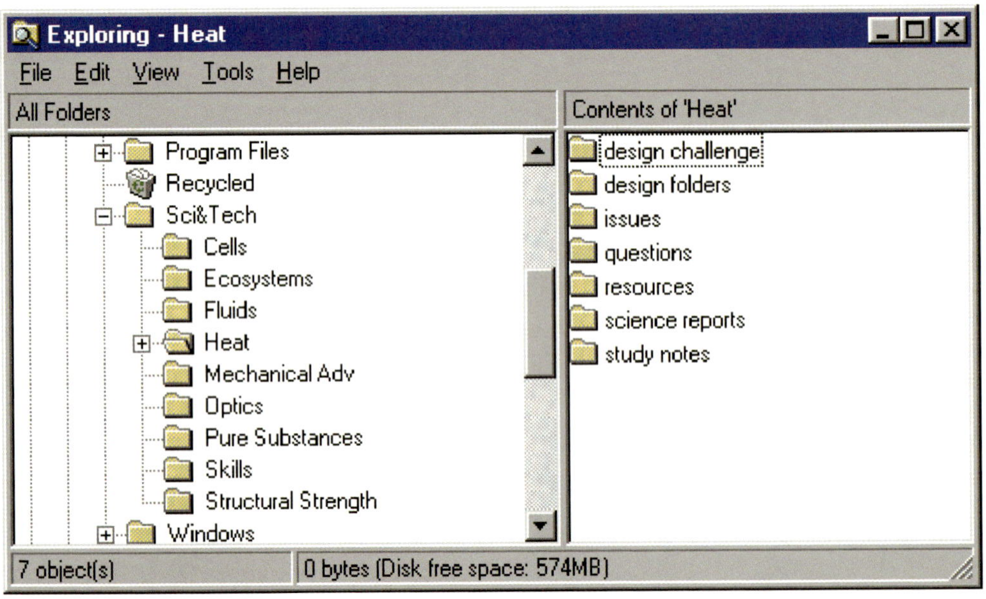

Using Spreadsheets and Databases

The computer is a powerful tool for storing, organizing, and analyzing data. Using databases and spreadsheets, you can record observations from science experiments or information acquired during the design process. You can use these software programs to sort information in different ways and to discover important relationships among variables. You can also use this software to create graphs and charts. The best way to get an overview of how spreadsheets and databases work is to explore different programs by using their built-in tutorials.

Spreadsheets

You can think of a spreadsheet as an electronic data table. To be as useful as possible, it has to be well designed. When you create any kind of data table, you should start by thinking carefully about what information you need to record, and how you will ultimately use that information.

In scientific inquiry, your spreadsheet will often be built on the variables you are using. If you are observing the viscosity of different fluids, your variables are the kind of fluid and the amount of time it takes for an object (e.g., a marble) to fall to the bottom of the receptacle containing that fluid (see **Figure 2**).

Figure 2

E7 =SUM(B7:D7)/3

	A	B	C	D	E
1	**Viscosity of Various Liquids**				
2		**Time taken for marble to fall through the liquid**			
3	**Liquid**	**Trial #1**	**Trail #2**	**Trial #3**	**Average**
4	water	0.48	0.48	0.47	0.48
5	canola oil	0.57	0.56	0.54	0.56
6	olive oil	0.45	0.42	0.47	0.45
7	corn syrup	0.52	0.54	0.55	0.54
8					

Your spreadsheet, like any data table, should have a title. The rows and columns of your spreadsheet should be well-labelled. Specific data is entered where a given row and column intersect. This space, which you can identify by referencing the number of the row and the letter of the column, is known as a cell. Some information you will enter yourself as numbers or text. In other cells, however, you will enter a formula that tells the program to perform a mathematical function using data in other cells. For instance, in **Figure 2**, you do not type in the numerical averages—the program does this for you based on the formula you have entered. To find out what formula is required, simply use the help function of the program you are using.

As you change or add data, the results of the calculations will change. The information in your spreadsheet will also be available as graphs—and these are also automatically updated as data is changed (**Figure 3**.)

Figure 3

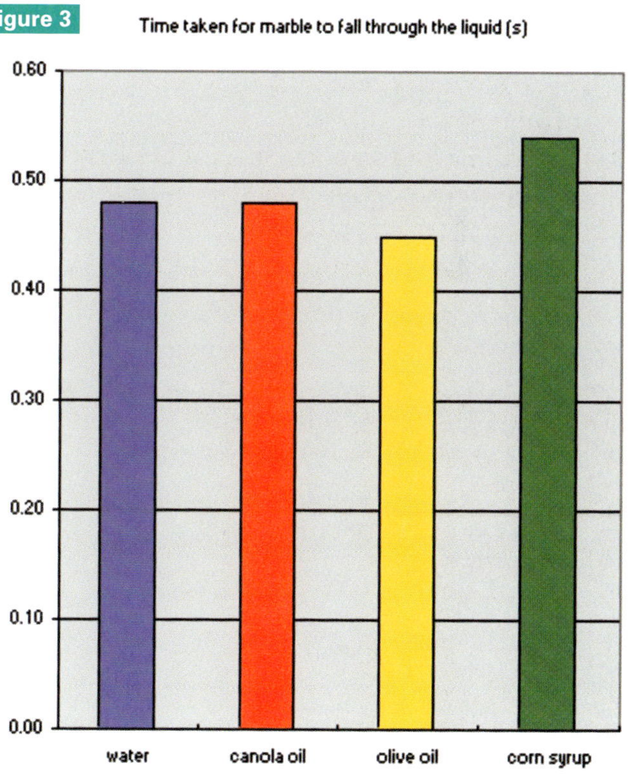

Databases

A database is another powerful tool for managing data. A good database will help you find specific information with ease, see relationships between different pieces of information, chart trends, create graphs, charts and reports, etc. Again, however, you have to spend some time designing this tool for your specific needs. A database can only give you what you need if you have designed it well.

First, you will need to become familiar with the power and uses of a database. The tutorials that are packaged with database software are good places to start. In general, a database is organized into records and fields. You can think of a record as an index card that lists all the information you need about a specific thing, person, or process. For instance, if you have conducted a survey about public reaction to the use of CFCs, each record would contain the information provided by one respondent.

Each record in a database is set up in the same way. The structure of each record is a "form" (**Figure 4**). You fill in the form with information you have gathered. Each piece of information is entered into a "field." Like spreadsheets, these fields have specific "electronic addresses." In fact, you can view information in a database as a list (**Figure 5**).

You can use the database for sophisticated analysis of your results. For instance, you can look for trends in responses to certain survey questions based on age, region, gender, etc. You can also use the data for other purposes (e.g., to create a mailing list). And, of course, you can generate accurate graphs and charts from the data.

Try This — Use a Database

Complete the tutorial for a popular database program. Now think of one project that you are working on now, or have recently completed, that would be enhanced by using a database. Design a form for that project.

Figure 4
A form designed for aircraft parts inventory

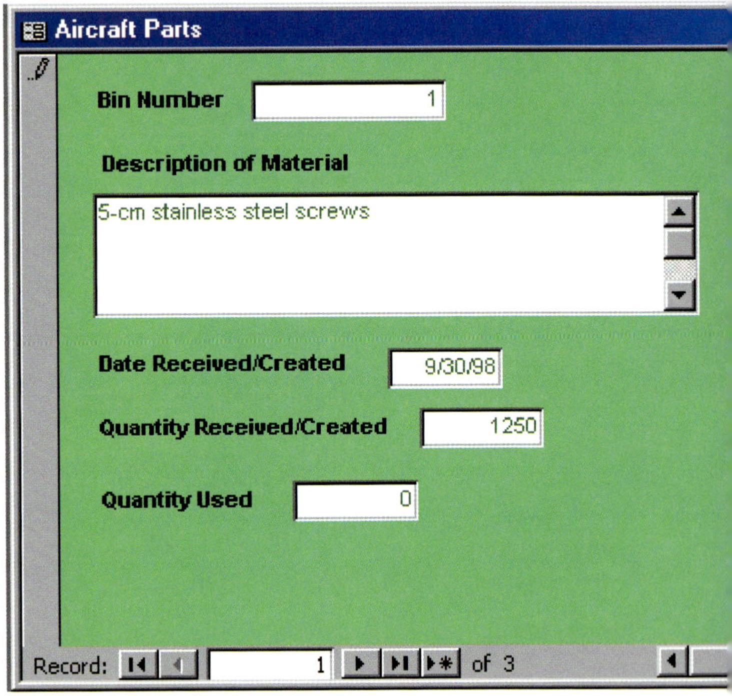

Figure 5
Inventory database in list view, showing 3 records

Aircraft Parts : Table

Bin Number	Description of Material	Date Received	Quantity Received	Quantity Used
1	5-cm stainless steel screws	9/30/98	1250	0
2	5-cm stainless steel nuts	9/30/98	1250	0
3	wheels	11/15/98	12	0

9D Working Together

Technological and scientific progress is almost always made by teams of people working together. Scientists share ideas, help each other design experiments and studies, and sharpen each other's conclusions. Technologists most frequently work in interdisciplinary teams, in which each person brings different expertise to the problem-solving process. We have all worked in a group at one time or another. In the "real world," group work is necessary and usually more productive than working alone. It is therefore important for us to be able to work in teams.

Try This The Barge Contest

Your group will compete against other design groups to create the barge capable of carrying the greatest number of marbles.

To solve this problem, you will build a model. Models are used as small-scale tests for larger, more expensive future experiments. Cheap models also allow a trial-and-error approach that would be far too expensive for full-scale tests.

Materials

- water
- large bucket
- marbles
- 250 g of modelling clay

Procedure

1. Working with all the members of your group, create a diagram for the barge's design. Everyone in the group must agree on this design (**Figure 1**).
2. Build your barge using only the modelling clay you have been given.
3. Fill the bucket with water and place your barge on the surface of the water.
4. Add marbles to your barge, until it begins to sink (**Figure 2**).
5. Record the maximum number of marbles that could be added.
6. Evaluate your teamwork.

Figure 1

Figure 2

Teamwork Tips

While working with *Nelson Science & Technology 7/8*, you will spend much of your time working in teams. Whenever you do, it is useful to take the following tips into account:

- Encourage all members to contribute to the work of the group.
- Respect everyone's contributions. There are many points of view and all perspectives should be considered. Keep an open mind.
- Be prepared to compromise.
- Share the work fairly.
- Keep focused on the task at hand. Divide the various tasks between all group members.
- Support the team's final decision.
- When you are given the opportunity of picking your own team, be sure that the students you decide to work with complement your strengths and weaknesses.

Can you think of any more tips that should be considered during group work?

Evaluating Teamwork

After you've completed a task with your team, evaluate the team's effectiveness—the strengths, weaknesses, opportunities, and challenges. Answer the following questions:

1. What were the strengths of your teamwork?
2. What were the weaknesses of your teamwork?
3. What opportunities were provided by working with your team?
4. What possible challenges did you see with respect to your teamwork?

9E Graphic Organizers

When you are trying to describe objects and events, it is sometimes helpful to write your ideas down so that you can see them and compare them with those of other people. Instead of writing these ideas down in sentences, they can often be represented visually. There are a number of ways of organizing information graphically that may help you express concepts and relationships.

Concept Maps

A concept map is a collection of words (representing concepts) or pictures that you connect to each other with arrows and short descriptions. The map is a drawing of what is happening in your brain.

You may have seen concept maps similar to the one in **Figure 1**. This concept map, called a food web, shows one way of thinking about animals and what they eat. The arrowhead points to the animal that eats the animal or plant at the other end of the arrow. The arrows describe the relationship between the organisms.

Concept maps can also be drawn of other topics and in other ways. For instance, the relationships between family members can be drawn using a concept map like the one in **Figure 2**.

Figure 2
A family

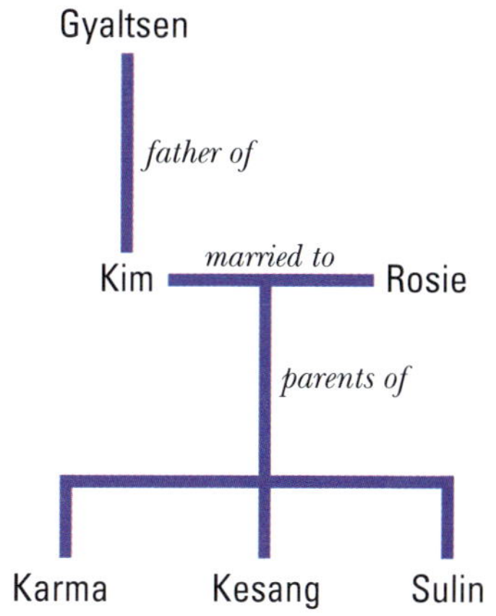

Figure 1
A food web

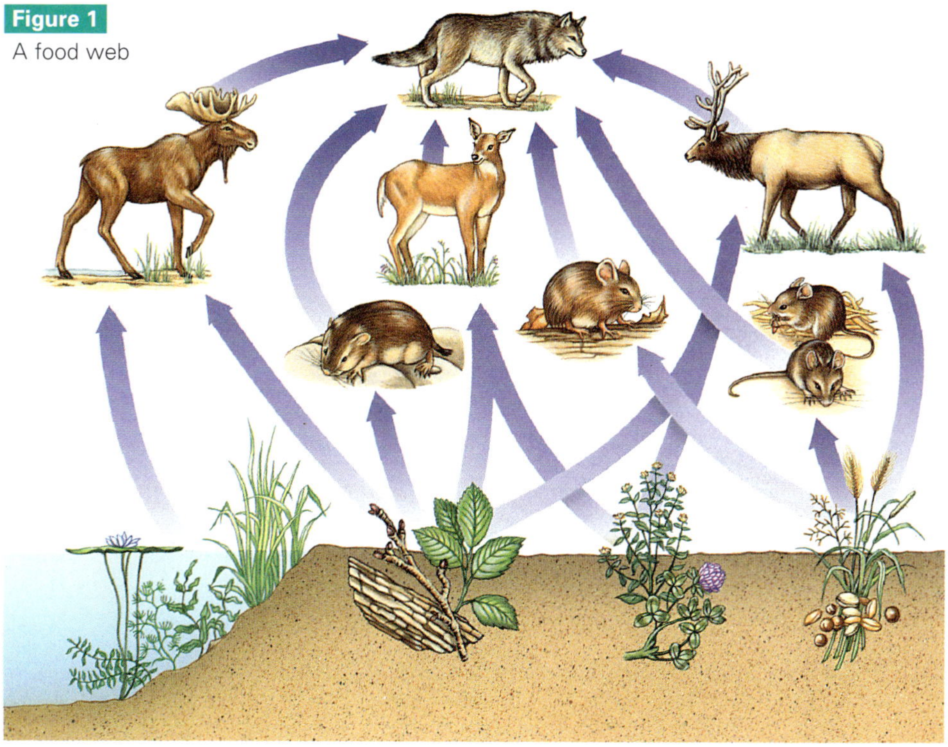

Concept maps can also be drawn to show a series of cause-and-effect relationships (**Figure 3**). You may find making this kind of map useful during the processes of scientific inquiry and design.

Sometimes concept maps can be used to show how your ideas change and become more complex as you work on a topic. You can see an example of this type of concept map in **Figure 4**.

Figure 3
Causes and effects related to plant growth

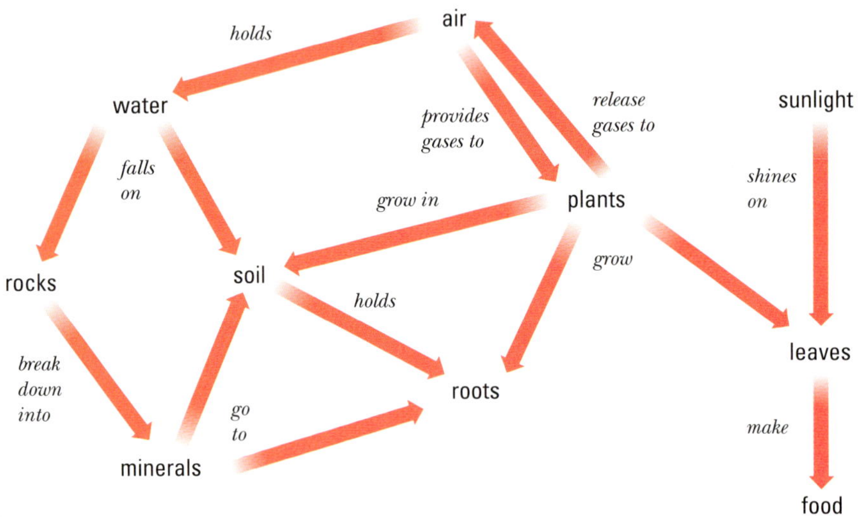

Figure 4
Concept maps evolve as your ideas evolve

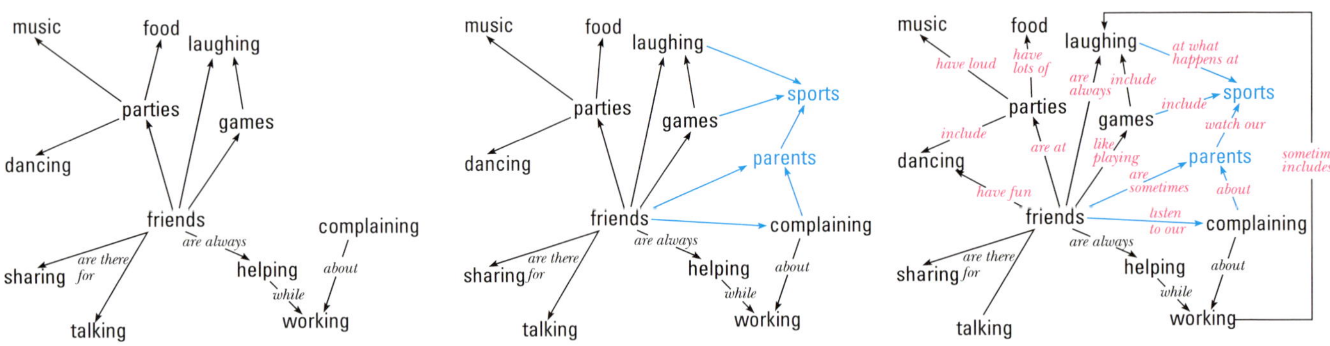

Making a Concept Map

Here are some steps you can take to help you make a concept map.

1. Choose the central idea of your concept map.
2. Write the central idea and all related ideas on small scraps of paper.
3. Move the scraps of paper around the central idea so that the ones most related to each other are close to each other. Ask yourself how they are related, and then use that information in the next steps.
4. On a big sheet of paper, write down all of your ideas in the same pattern that you have arranged the scraps. Draw arrows between the ideas that are related.
5. On each arrow, write a short description of how the terms are related to each other.

As you go, you may find other ideas or relationships. Add them to the map. When you gain new ideas—whether from research, from your investigations, or from other people—go ahead and change your concept map. You may want to add new ideas in a different colour of ink, to indicate your new ways of thinking about the ideas.

Comparison Matrix

It is often useful to use a comparison matrix to compare things according to a variety of criteria (**Table 1**). For instance, you might want to compare the properties of different materials in order to choose the best one for a particular design.

Table 1 The Heating Curve of Water

	Properties			
	Malleable	**Lustrous**	**Strong**	**Affordable**
Gold	X	X		
Silver	X	X	X	
Copper	X	X	X	X

Venn Diagram

Another simple diagram used for comparing and contrasting things is the Venn diagram (**Figure 5**). Similarities between two items are written in the area of overlap; differences are written in the separate circles.

Figure 5

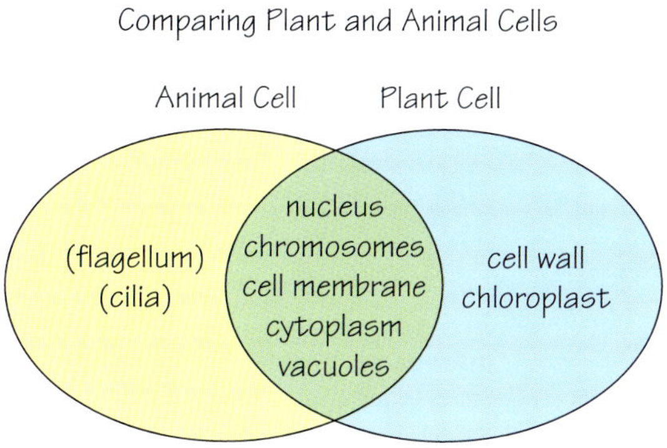

Hierarchical (Branching) Diagram

Some concepts can be broken down into subcategories. A hierarchical diagram is a useful way of representing these relationships (**Figure 6**).

Figure 6

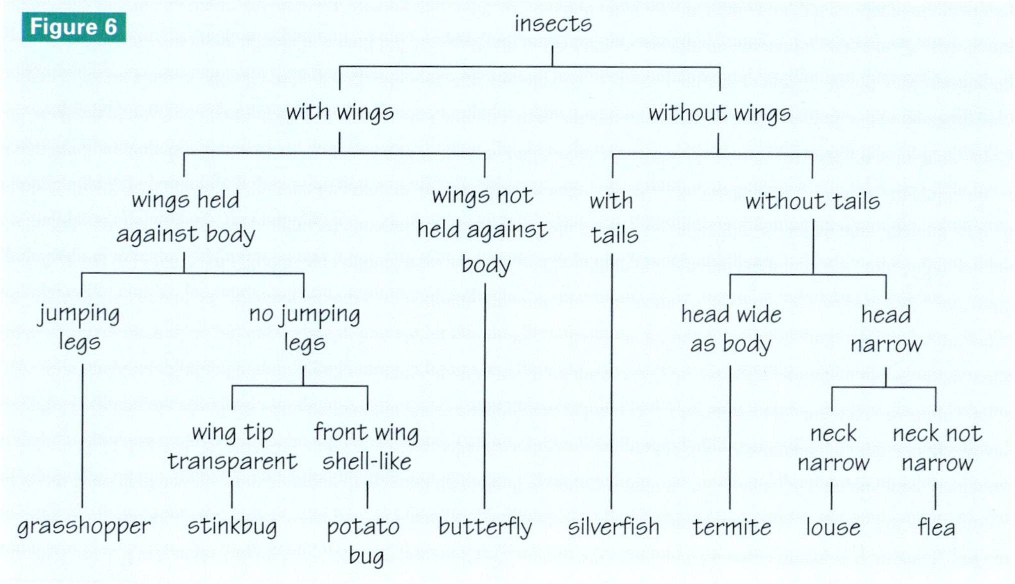

Target Diagram

A target diagram is often used to illustrate which item best applies to a given criterion. (**Figure 7**).

Figure 7

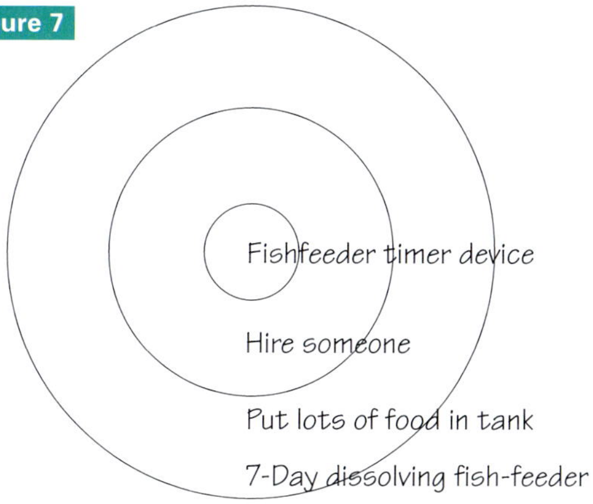

Photo Credits

Cover: collage of (clockwise from top left) Corel, Alan Marsh/First Light, Visuals Unlimited/Derrick Ditchburn, B. Fraunfelter/First Light; p. 2 CORBIS/Rodney Hyett/Elizabeth Whiting & Associates; p. 3 Paul Till; p. 5 top Ian Crysler, bottom Dave Starrett; pp. 6–8 Dave Starrett; p. 9 Paul Till; p. 12 Dave Starrett; pp. 13–16 Ian Crysler; pp. 17–19 Dave Starrett; p. 20 PhotoDisc; p. 22 Visuals Unlimited/Jeff Greenberg; p. 23 CORBIS/Richard T. Nowitz; p. 24 Dave Starrett; p. 26 Dave Starrett; p. 30 top to bottom PhotoDisc, Todd Ryoji, Dick Hemingway, Omni Photo Communications Inc./T. W. Image Network; p. 32 Dave Starrett; pp. 34–35 Dave Starrett; p. 37 Dave Starrett; p. 40 top Visuals Unlimited/Richard L. Carlton, rest Boréal; p. 41 top left Ian Crysler, bottom right Boréal; p. 49 Dave Starrett; p. 52 top left CORBIS/Rodney Hyett/Elizabeth Whiting & Associates, top right R. W. Jones/First Light, middle right (b) Bob Semple, (c) First Light, bottom right Ron Watts/First Light; p. 53 top left Visuals Unlimited/R. Lindholm, middle left Joe Lepiano/Artbase, bottom left Dave Starrett, top right Dave Starrett, bottom right Andrea Pistolesi/The Image Bank; p. 54 top left Dave Starrett, bottom left Gary Russ/The Image Bank; p. 57 left Dave Starrett, top right Joe Lepiano/Artbase, middle right Dave Starrett, bottom right Chuck Savage/First Light; p. 58 Paul Till; pp. 73–75 Dave Starrett; pp. 86–87 Dave Starrett; p. 91 Ian Crysler.